Materials Forming, Machining and Tribology

Series editor

J. Paulo Davim, Aveiro, Portugal

More information about this series at http://www.springer.com/series/11181

Kaushik Kumar · Divya Zindani
Nisha Kumari · J. Paulo Davim
Editors

Micro and Nano Machining of Engineering Materials

Recent Developments

Editors
Kaushik Kumar
Department of Mechanical Engineering
Birla Institute of Technology, Mesra
Ranchi, Jharkhand, India

Nisha Kumari
Department of Mechanical Engineering
Birla Institute of Technology, Mesra
Ranchi, Jharkhand, India

Divya Zindani
Department of Mechanical Engineering
National Institute of Technology, Silchar
Cachar, Assam, India

J. Paulo Davim
Department of Mechanical Engineering
University of Aveiro
Aveiro, Portugal

ISSN 2195-0911 ISSN 2195-092X (electronic)
Materials Forming, Machining and Tribology
ISBN 978-3-030-07642-9 ISBN 978-3-319-99900-5 (eBook)
https://doi.org/10.1007/978-3-319-99900-5

This Springer imprint is published by the registered company Springer Nature Switzerland AG
The registered company address is: Gewerbestrasse 11, 6330 Cham, Switzerland

Preface

In the present period of 4th Industrial Revolution "Industry 4.0" (I40) which calls for analog to digital transformation of the entire industrial production utilizing amalgamation of manufacturing sector with Internet and Information and Communication Technologies (ICT). All the focus now has been primarily directed toward customer satisfaction, which calls for optimal cost, time, and quality. Aimed toward these three variables, companies are striving continuously to develop/improve practices and techniques to fulfill consumer requirements and in turn increase their market share and profit.

Manufacturing sectors till last decades, for profit and survival, completely dedicated their energy and expertise toward mass manufacturing for most of their work. But with the evolution of new era of digitalization and customization, industries are changing their focus from mass to customized manufacturing and in this transformation, one aspect is **Micro and Nano machining**. Often, people get confused with the basic ideology behind micromachining and they consider it to be "machining of highly miniature components with miniature features" which is quite inappropriate and it can be correctly defined as "material removal at micro/nano level with no constraint on the size of the component" or in broader terms "material removal at micro/nano level for macro or micro/nano components."

In recent years, there has been a rapid growth in the demand to produce micro/nano size products to cater to the needs of the industry. This has resulted toward a miniaturization drive. The miniature sized products are being used in large number of different engineering disciplines because of their unique small-sized characteristics. Micro-engines, micro-heat exchangers, micro-pumps, micro-channels, printing heads, and medical implants are some of the typical applications of micro and nano sized products. These micro system-based products are an important contributor to a sustainable economy as they represent a key to survival for many companies.

To meet the ever growing demand for the miniature products, studies have been carried out on developing different micro and nano manufacturing technologies. Recent research focuses on to develop new micro and nano manufacturing platforms while integrating the different technologies to manufacture the micro and nano components in a high-throughput and cost-effective manner.

Manufacturing processes at the microscale are the key-enabling technologies to bridge the gap between the nano- and the macro-worlds to increase the accuracy of micro/nano-precision production technologies, and to integrate different dimensional scales in mass manufacturing processes. Accordingly, this edited volume is a result of rapid dissemination of original theoretical and applied research in the areas of micro- and nano-manufacturing that are related to process innovation, accuracy, and precision, throughput enhancement, material utilization, compact equipment development, environmental and life cycle analysis, and predictive modeling of manufacturing processes with feature sizes less than one hundred micrometers.

The main objective of the book is dedicated to Micro and Nano Machining and is targeted to cater the needs of all academics students, researchers and industry practitioners, engineers, and research scientists/academicians involved in machining at micro/nano level toward creating Macro or Micro sized components.

The chapters in the book have been categorized in **three parts**, namely, **Part I: Overview of Micro and Nano Machining; Part II: Micro and Nano Machining with Conventional Machining Techniques; and Part III: Micro and Nano Machining with Nonconventional Machining Techniques.**

Part I contains Chapter “Recent Trends in Micro and Nano Machining of Engineering Materials” and Chapter “Micro and Nano Machining—An Industrial Perspective”, whereas Part II has Chapter “Micromachining of Titanium Alloys” and Chapter “Ultra-Precision Diamond Turning Process” and Part III with Chapters “Abrasive Waterjet Cutting of Lanthanum Phosphate—Yttria Composite: A Comparative Approach”–“Experimental Analysis of Wire EDM Process Parameters for Micro Machining of High Carbon High Chromium Steel by Using MOORA Technique”.

Part I starts with Chapter “Recent Trends in Micro and Nano Machining of Engineering Materials” which discusses the recent trends of micro and nano machining of components with an emphasis on nonconventional technologies. The chapter describes the various techniques, i.e., electrical discharge machining (EDM), electrochemical machining (ECM), abrasive water-jet machining (AWJM), and laser beam machining (LBM). A brief section on recent trends in nano machining has also been presented at the end of the chapter.

Chapter “Micro and Nano Machining—An Industrial Perspective” provides an insight into micro and nano-machining with an industrial perspective. The applications to important and current fields like biotechnology, electronics, medicine, optics, aviation, automobile, and communication have been depicted. So this chapter describes micro and nano machining techniques, which are quite sophisticated and complicated when compared to their conventional or traditional part, with their applications in industry.

Chapter “Micromachining of Titanium Alloys”, the first chapter of Part II, enlightens the readers with micromachining of titanium alloys. Titanium alloys are widely used in aerospace, automotive, and biomedical industries due to their high specific strength (strength-to-weight ratio), excellent mechanical properties, outstanding corrosion resistance, and biocompatibility. The biocompatibility has made it an evident choice as materials for applications like next-generation vascular stents, drug-eluting stents (DES), micro-opto-electromechanical systems (MOEMS), microfluidics, and bio-microelectromechanical systems (bio-MEMS). In this chapter, a comprehensive overview on micromachining of titanium alloys has been presented utilizing the available literature on conventional, nonconventional, and hybrid micromachining of various titanium alloys. The chapter also describes effects of key machining parameters and performance criterions depending on previous works on those respective fields. In addition, micromachining challenges, such as burr formation, tool wear, property change of recast layer, improper surface finish, etc., have been addressed for each machining process, and probable steps to overcome those issues have been suggested. In the end, future research direction on the domain of micromachining of titanium alloys has also been discussed.

Chapter "Ultra-Precision Diamond Turning Process" describes ultra-precision diamond turning process. Ultra-precision machining is the efficient technique to produce highly precise surface with complex shapes and micro features. Surface quality in ultra-precision machining is only assured by strictly following the optimized process conditions. This chapter covers the various stages of the ultra-precision diamond turning process especially considering the practical aspects. This chapter also provides the understanding of, importance and effects of each stage of ultra-precision machining including metrology.

Chapter "Abrasive Waterjet Cutting of Lanthanum Phosphate—Yttria Composite: A Comparative Approach", which commences Part III, concentrates on application of abrasive water-jet cutting (AWJC), a nonconventional machining technique, for micromachining of lanthanum phosphate—yttria composite. The objective of the chapter is to elaborate on the effect of silicon carbide (SiC) and garnet of each 80 mesh size, used as abrasives, on micromachining of $LaPOd_4$ + $20\%Y_2O_3$ composite. The output responses, material removal rate (MRR), kerf angle (KA), and surface roughness (Ra) are measured with varied input parameters, jet pressure (JP), stand-off distance (SOD), and traverse speed (TS). The finding of the chapter provides a comparison of the performance of two abrasive materials toward indicated outputs. Microscopy examination was also performed to elaborate on the failure mode. The chapter concludes with the factual decision of the choice of abrasive for better responses.

In Chapter "Laser Micromachining of Engineering Materials—A Review", review of laser micromachining of engineering materials has been depicted. Laser micromachining is a precise nonconventional type of machining process which is used in the fabrication of micro-components ranging up to 500μm. Laser ablation distinctively focuses on the small elemental areas, which helps in absorbing high percentage of energy and results in machining at micron level. The chapter represents an overview of various researches carried out in laser machining fields, its applicability, and the advancements made. It also shows the implementation of ultra-short and femtosecond pulsed laser micromachining. Femtosecond laser machining can machine even brittle and transparent materials like glass, sapphire, etc. The performance and appreciation of laser beam micromachining (LBMM) on some of the advanced engineering materials has been elaborately discussed in this chapter.

Chapter "Experimental Analysis of Wire EDM Process Parameters for Micro Machining of High Carbon High Chromium Steel by Using MOORA Technique", the concluding chapter of the section and the book, concentrates on experimentation on wire EDM, another very important nonconventional machining technique, toward micro machining of high carbon high chromium steel. The chapter further provides the optimal process parameters for best outputs using a multi-criteria decision-making optimization technique (MCDM), MOORA technique. The chapter considers input parameters like pulse width time, wire feed rate, and pulse off time on responses, namely, material removal rate, surface roughness, and kerf width of the high carbon and high chromium steel during micromachining using wire electrical discharge machining process. In order to find the optimal solution multi-objective optimization using an MCDM technique, i.e., MOORA was applied. Apart from finding and validating the optimal process parameters, the chapter also indicated that in future, the outcome of other input process parameters like wire material, wire tension, flushing pressure, and dielectric fluid type can be investigated to attain enhanced material removal rate, better surface quality, and cutting of nonconducting materials.

First and foremost, the Editors would like to thank God. It was His blessings that this work could be completed to their satisfaction. You have given the power to believe in passion, hard work, and pursue dreams. The Editors could never have done this herculean task without the faith they have in you, the Almighty. They are thankful for this.

The Editors would also like to thank all the Chapter Contributors, the Reviewers, the Editorial Advisory Board Members, Book Development Editor, and the team of Publisher Springer Nature for their availability for work on this editorial project.

Throughout the process of editing this book, many individuals, from different walks of life, have taken time out to help. Last but definitely not least, the Editors would like to thank them all, their well-wishers, for providing them encouragement. They would have probably given up without their support.

Ranchi, India — Kaushik Kumar
Cachar, India — Divya Zindani
Ranchi, India — Nisha Kumari
Aveiro, Portugal — J. Paulo Davim

Contents

About the Editors

Dr. Kaushik Kumar B.Tech. (Mechanical Engineering, REC (Now NIT), Warangal), MBA (Marketing, IGNOU) and Ph.D. (Engineering, Jadavpur University), is presently an Associate Professor in the Department of Mechanical Engineering, Birla Institute of Technology, Mesra, Ranchi, India. He has 16 years of Teaching and Research and over 11 years of industrial experience in a manufacturing unit of global repute. His areas of teaching and research interest are Conventional and Nonconventional Quality Management Systems, Optimization, Nonconventional machining, CAD / CAM, Rapid Prototyping, and Composites. He has 9 Patents, 16 Book, 13 Edited Book, 38 Book Chapters, 122 International Journal Publications, and 21 International and 8 National Conference publications to his credit. He is on the editorial board and review panel of 7 International and 1 National Journals of repute. He has been felicitated with many awards and honors.

Divya Zindani B.E. (Mechanical Engineering, Rajasthan Technical University, Kota), M.E. (Design of Mechanical Equipment, BIT Mesra), presently pursuing Ph.D. (National Institute of Technology, Silchar). He has over 2 years of Industrial experience. His areas of interests are Optimization, Product and Process Design, CAD/CAM/CAE, Rapid Prototyping, and Material Selection. He has 1 Patent, 4 Books, 6 Edited Books, 18 Book Chapters, 2 SCI Journal, 7 Scopus Indexed International Journal, and 4 International Conference Publications to his credit.

Ms. Nisha Kumari B.E. (Mechanical Engineering, Institute of Technical Education and Research, Bhubaneswar) and M.E. (Design of Mechanical Equipment, BIT Mesra). She has over 1 year of Industrial experience. Her areas of interests are Biomechanics, Product and Process Design, CAD/CAM/CAE, and Rapid Prototyping. She has 2 Edited Books, 2 Book, 4 Book Chapters, 5 International Journal, and 1 National Conference Publications to her credit. She has been felicitated with awards and honors.

Prof. J. Paulo Davim received his Ph.D. in Mechanical Engineering in 1997, M.Sc. in Mechanical Engineering (Materials and Manufacturing Processes) in 1991, Licentiate degree (5 years) in Mechanical Engineering in 1986 from the University of Porto (FEUP), the Aggregate title from the University of Coimbra in 2005, and D.Sc. from London Metropolitan University in 2013. He is EUR ING by FEANI and senior chartered engineer by the Portuguese Institution of Engineers with MBA and specialist title in engineering and industrial management. Currently, he is Professor at the Department of Mechanical Engineering of the University of Aveiro. He has more than 30 years of teaching and research experience in manufacturing, materials, and mechanical engineering with special emphasis on machining and tribology. Recently, he has also interest in management/industrial engineering and higher education for sustainability/engineering education. He has received several scientific awards. He has worked as evaluator of projects for international research agencies as well as examiner of Ph.D. thesis for many universities. He is the editor in chief of several international journals, guest editor of journals, editor of books, series editor of books, and scientific advisor for many international journals and conferences. At present, he is an editorial board member of 25 international journals and acts as reviewer for more than 80 prestigious Web of Science journals. In addition, he has also published as editor (and co-editor) for more than 100 books and as author (and co-author) for more than 10 books, 80 book chapters, and 400 articles in journals and conferences (more than 200 articles in journals indexed in Web of Science/h-index 45+ and SCOPUS/h-index 52+).

Part I
Overview of Micro and Nano Machining

Recent Trends in Micro and Nano Machining of Engineering Materials

T. Muthuramalingam

1 Introduction

The utilization of higher strength materials such as titanium alloy, nickel-based alloy, and stainless steel in the modern engineering fields has grown rapidly. However, the machining of such materials is a complex one [1–3]. The surface quality of the machined material is mainly determined by the volume of material removal. If the material removal is in micron level, the surface roughness is reduced considerably. This kind of material removal can be termed as micro machining. Hence, the lifetime of the product can be considerably increased. The lifetime can be further enhanced by the nano machining process. Nevertheless, it is very tedious to remove the material either in micro or nano level from the high strength engineering materials using conventional machining process. Hence, the unconventional machining processes such as electrical discharge machining (EDM), electrochemical machining (ECM), abrasive water-jet machining (AWJM), and laser beam machining (LBM) can be used to machine such materials [4, 5].

2 Micro Machining with Electrical Discharge Machining Process

Electrical discharge machining (EDM) is one of the unconventional machining processes in which electrical energy is applied between tool and workpiece under pulse form. This generates enormous heat which is capable to melt and vaporize the material. The melted material is resolidified for making the white layer thickness

T. Muthuramalingam (✉)
Department of Mechatronics Engineering, School of Mechanical Engineering, SRM Institute of Science and Technology, Kattankulathur Campus, Chennai 603203, India
e-mail: muthu1060@gmail.com

K. Kumar et al. (eds.), *Micro and Nano Machining of Engineering Materials*, Materials Forming, Machining and Tribology,
https://doi.org/10.1007/978-3-319-99900-5_1

over the machined surface. The nature of the pulse generator has high influences on determining surface quality of the material. In micro EDM, the electrical pulses are applied with the pulse duration of the micro seconds. The RC pulse generator produces considerable better surface quality whereas transistor pulse generator produces a higher material removal rate. The iso energy pulse generator can produce better surface quality and higher material removal rate than the existing pulse generators as shown in Fig. 1. A lot of research attention has been given to analyze the effects of electrical process parameters in micro EDM process.

In wire EDM process, the tool electrode is replaced by wire electrode for the machining purposes. The micro machining is possible by applying smaller energy across the machining zone [6]. The surface hardness of the machined material using EDM process is mainly influenced by the white layer thickness. Since micro machining can produce only tiny energy distribution, it can produce narrow white layer thickness. Titanium alloys are highly utilized in the manufacturing industries owing to its higher corrosive resistance and strength. It is proved titanium alloy (Ti-6Al-4V) can be effectively machined using WEDM process [7]. The machinability can be further increased by adopting multi-criteria decision-making algorithms [8]. The considerable amount of research works are being carried out to analyze the effects of the tool electrode size in micro EDM for enhancing the machining process. The positive polarity of the tool electrode can reduce the surface roughness considerably [9]. The micro EDM process can produce dies and moulds with intricate shapes using higher strength engineering materials easily. Nevertheless, micro EDM process has some demerits such as high machining time, heat affected zone, residual stress formation, and the ability to machine only electrical conducting materials [10]. Since the air gap determines the spark energy in micro EDM process, it has to be controlled in an effective manner. The monitoring and controlling of process parameters such air gap and peak current can enhance the performance measures in micro EDM process [11–16].

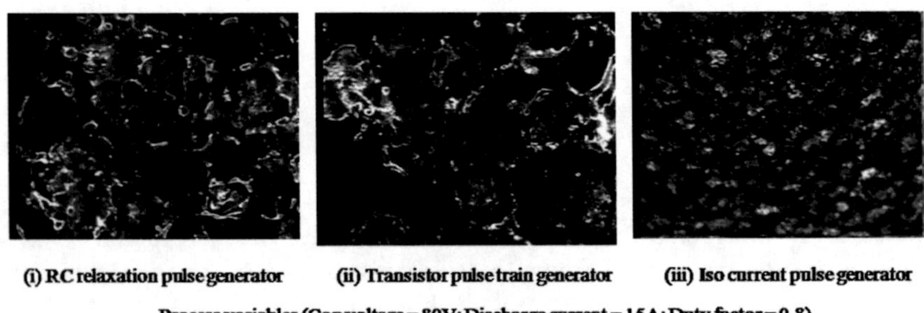

Fig. 1 Effects of pulse generators on surface topography

3 Micro Machining with Electro Chemical Machining Process

Electrochemical micro machining (ECMM) is one of the unconventional machining processes in which the electrical energy is applied between tool and workpiece under electrolyte environment. This removes the material by anodic dissolution of the workpiece material during the machining process. The optimal selection of electrolyte medium can enhance the machinability of high strength electrical conducting nickel-based alloy materials. The proper coating of tool electrodes such as nickel and chromium coating enhances the performance measures in ECMM process. The optimal combination of process parameters helps to obtain better machining characteristics in ECMM process. Various multiple response optimization techniques include Taguchi Grey analysis, Taguchi-DEAR method, TOPSIS, and neural networks. The controlling of process variables can enhance the surface morphology of machined materials in ECMM process. Figure 2 shows the surface morphology of the machined Inconel alloy in ECMM process using scanning electron microscope. ECMM process has the advantages of high machining rate and no tool wear. However, the cracks may be produced owing to the residual stress formation [17–20].

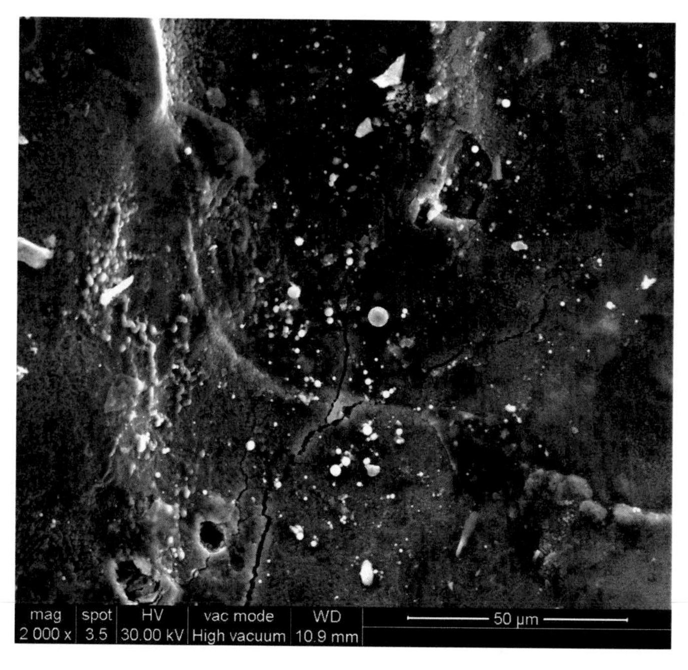

Fig. 2 Surface morphology of the machined Inconel alloy using ECMM

4 Micro Machining with Abrasive Water Jet Machining Process

Abrasive water-jet machining process (AWJM) is an important unconventional machining process in which the abrasive particles mixed pressurized water is made to contact with the workpiece. Normally garnet, silicon carbide, and boron nitride can be used as the abrasive particles in AWJM process. Unlike EDM and ECM process, AWJM process can machine the nonelectrical conductive materials with less heat affected zone and no recast layer thickness [4]. However, the taperness is unavoidable effect in AWJM process. Most attention has to be given for reducing the taperness as much as possible. The usage of metal matrix composites (MMC) is increased due to its distinct physical properties. The MMC such as aluminum MMC can be easily machined using this process. Since the size of the particles, standoff distance, and flow rate of jet can affect the performance measures of the machining process, it is important to choose optimal process parameters in AWJM process. The micron level of material removal in AWJM is possible by producing lower impact energy. This can be obtained by applying optimal lower process variables by incorporating control algorithms.

5 Micro Machining with Laser Beam Machining Process

Traditionally, conventional machining processes require high strength tool design than the workpiece. These processes are also resulted in poor surface finish of the workpiece at the micro level due to abrasion. Laser beam machining (LBM) is not only coherent but also monochromatic in nature. This concentrated energy is so intense that it has the capability of melting diamond within a fraction of seconds. Laser tries to propagate through the workpiece and the energy of the photon gets absorbed by the workpiece and temperature rises rapidly, as a result, the material gets removed due to evaporation and melting. Laser beam machining is preferred over other unconventional machining because of its high material removal rate and very good surface finish. Meijer discussed laser beam generation, the effects of different pulse length, and wavelengths during the machining process [21]. The laser drilling process has been numerically simulated to analyze the effects of laser beam power on machinability [22–24]. It is essential to investigate the effects of process parameters on circularity for increasing machining accuracy. It has been observed that the LBM process can enhance the machinability of steel over the traditional machining process [25]. Taguchi methods for experiment design have been utilized in LBM process to increase the efficacy. It has been proved that the process parameters can influence the machinability in LBM process [26]. The modification of laser medium can also significantly affect the machining characteristics in LBM process [27].

6 Nano Machining with Unconventional Machining Process

The nano level of material removal can enhance the life cycle of the product than the micro machining process. Hence, it is important to introduce the nano machining process to obtain better product quality of engineering materials. The nano machining in EDM and WEDM process is possible by adding carbon nano tubes in the insulating medium with uniform distribution. In AWJM process, the nano level of material removal can be performed by mixing abrasive particles with nano size. Whereas the low power solid state laser module can remove the material under nano level of material removal in LBM process. The nano abrasive particles reinforced magneto rheological based nano machining process can provide better nano machining of the engineering materials.

References

1. K.J. Ashwin, T. Muthuramalingam, Influence of duty factor of pulse generator in electrical discharge machining process. Int. J. Appl. Eng. Res. **12**(21), 11397–11399 (2017)
2. T. Muthuramalingam, S. Vasanth, T. Geethapriyan, Influence of energy distribution and process parameters on tool wear in electrical discharge machining. Int. J. Control Theory Appl. **9**(37), 353–359 (2016)
3. T. Geethapriyan, K. Kalaichelvan, T. Muthuramalingam, Multi performance optimization of electrochemical micro-machining process surface related parameters on machining Inconel 718 using Taguchi-grey relational analysis. La Metal. Ital. **2016**(4), 13–19 (2016)
4. S. Vasanth, T. Muthuramalingam, P. Vinothkumar, T. Geethapriyan, G. Murali, Performance analysis of process parameters on machining titanium (Ti-6Al-4V) alloy using abrasive water jet machining process. Proc. CIRP **46**(1), 139–142 (2016)
5. T. Geethapriyan, K. Kalaichelvan, T. Muthuramalingam, Influence of coated tool electrode on drilling Inconel alloy 718 in electrochemical micro machining. Proc. CIRP **46**(1), 127–130 (2016)
6. A. Ramamurthy, R. Sivaramakrishnan, T. Muthuramalingam, S. Venugopal, Performance analysis of wire electrodes on machining Ti-6Al-4V alloy using wire electrical discharge machining process. Mach. Sci. Technol. **19**(4), 577–593 (2015)
7. A. Ramamurthy, R. Sivaramakrishnan, T. Muthuramalingam, Taguchi-grey computation methodology for optimum multiple performance measures on machining titanium alloy in WEDM process. Ind. J. Eng. Mat. Sci. **22**(2), 181–186 (2015)
8. T. Muthuramalingam, B. Mohan, Application of Taguchi-grey multi responses optimization on process parameters in electro erosion. Measurement **54**(1), 495–502 (2014)
9. T. Muthuramalingam, B. Mohan, A. Jothilingam, Effect of tool electrode re-solidification on surface hardness in electrical discharge machining. Mater. Manuf. Process. **29**(11–12), 1374–1380 (2014)
10. T. Muthuramalingam, B. Mohan, A review on influence of electrical process parameters in EDM process. Arch. Civil Mech. Eng. **15**(1), 87–94 (2014)
11. T. Muthuramalingam, B. Mohan, A. Rajadurai, D. Saravanakumar, Monitoring and fuzzy control approach for efficient electrical discharge machining process. Mater. Manuf. Process. **29**(3), 281–286 (2014)

12. T. Muthuramalingam, B. Mohan, Performance analysis of iso current pulse generator on machining characteristics in EDM process. Arch. Civil Mech. Eng. **14**(3), 383–390 (2014)
13. T. Muthuramalingam, B. Mohan, A. Rajadurai, M.D. Antony, Experimental investigation of iso energy pulse generator on performance measures in EDM. Mater. Manuf. Process. **28**(10), 1137–1142 (2013)
14. T. Muthuramalingam, B. Mohan, Influence of tool electrode properties on machinability in electrical discharge machining. Mater. Manuf. Process. **28**(8), 939–943 (2013)
15. T. Muthuramalingam, B. Mohan, Influence of discharge current pulse on machinability in electrical discharge machining. Mater. Manuf. Processes **28**(4), 375–380 (2013)
16. T. Muthuramalingam, B. Mohan, Design and fabrication of control system based iso current pulse generator for electrical discharge machining. Int. J. Mech. Manuf. Syst. **6**(2), 133–143 (2013)
17. T. Muthuramalingam, B. Mohan, A study on improving machining characteristics of electrical discharge machining with modified transistor pulse generator. Int. J. Manuf. Technol. Manage. **27**(1/2/3), 101–111 (2013)
18. T. Muthuramalingam, B. Mohan, Enhancing the surface quality by iso pulse generator in EDM process. Adv. Mater. Res. **622–623**(1), 380–384 (2013)
19. T. Muthuramalingam, B. Mohan, Taguchi-grey relational based multi response optimization of electrical process parameters in electrical discharge machining. Ind. J. Eng. Mat. Sci. **20**(6), 471–475 (2013)
20. T. Geethapriyan, K. Kalaichelvan, A. Rajadurai, T. Muthuramalingam, S. Naveen, A review on investigating the effects of process parameters in electrochemical machining. Int. J. Appl. Eng. Res. **10**(2), 1743–1748 (2015)
21. J. Meijer, Laser beam machining (LBM) state of the art and new opportunities. J. Mat. Process. Technol. **149**, 2–17 (2004)
22. D. Abidou, N. Yusoff, N. Nazri, M.A.O. Awang, M.A. Hassan, Numerical simulation of metal removal in laser drilling using symmetric smoothed particle hydrodynamics. Prec. Eng. **49**, 69–77 (2017)
23. P. Parandoush, A. Hossain, A review of modeling and simulation of laser beam machining. Int. J. Mach. Tools Manuf. **85**, 135–145 (2014)
24. N.S. Amalina, A. Hossain, K. Alrashed, Y. Nukmana, Prediction modelling of recast layer and circularity for laser drilling of polyethylene terephthalate (PETG) thermoplastic. Proc. Eng. **184**, 197–204 (2017)
25. M.J. Jackson, W.O. Neill, Laser micro-drilling of tool steel using Nd:YAG lasers. J. Mater. Process. Technol. **142**, 517–525 (2003)
26. A.K. Dubey, V. Yadava, Multi objective optimization of laser beam cutting process. Opt. Laser Technol. **40**, 562–570 (2007)
27. A.K. Dubey, V. Yadava, Laser beam machining-a review. Int. J. Mach. Tools Manuf. **49**, 609–628 (2008)

Micro and Nano Machining—An Industrial Perspective

Nadeem Faisal, Divya Zindani and Kaushik Kumar

1 Introduction

Micromachining comes from the combination of two words, "micro" and "machining", i.e., removal of material at the micro level. Often, people get confused with the basic ideology behind micromachining and they consider it to be "machining of highly miniature components with miniature features" which is quite incorrect and it can be correctly defined as "material removal process at micro/nano level with no constraint on the size of the component" or in broader terms "macro components but material removal rate is at micro/nano level or micro/nano components and material removal rate is at micro/nano level."

With global competitiveness and growing commercialization and the competition to conquer the market, the requirement and need of today's industry lead toward the continuous refinement of manufacturing processes as far as the achieved precision is concerned. Large-scale components such as precision molds, machine elements, mirrors, and lenses require micrometer or sometimes nanometre tolerances and surface finishes. At the same time, micro parts are introduced in several applications, whose functioning would otherwise be impossible [1, 2]. The most characteristic example is the integrated circuit (IC) industry, which creates parts that have micrometer dimensions or possess features with nanometre size [3]. Furthermore, microelectromechanical systems (MEMS), which usually contain micro-sized mechanical parts, are becoming more and more popular in engineering. The outcome is the generation of applications that are efficient, reliable, environmentally friendly, and more economical.

N. Faisal (✉) · K. Kumar
Department of Mechanical Engineering, BIT Mesra, Ranchi, India
e-mail: ndmfaisal@gmail.com

D. Zindani
Department of Mechanical Engineering, NIT Silchar, Silchar, India

K. Kumar et al. (eds.), *Micro and Nano Machining of Engineering Materials*, Materials Forming, Machining and Tribology,
https://doi.org/10.1007/978-3-319-99900-5_2

The few of the abovementioned trends are based on disciplines that are interdependent and need to be taken into account. The technologies required are either new ones developed purely for their application in this specific field or ones that are currently in use and are being further extended in order to accomplish the task. The realization of state-of-the-art miniature devices is achieved through highly advanced manufacturing processes, which are able to provide the required tolerance, roughness, and size. This is achieved by developing new machine tools and control systems and even wholly novel processes. This should be combined with techniques that can observe and measure features at nanometre and sub-nanometre levels and even manipulate atoms with sub-nanometre control, and new modeling and simulation techniques that are used for the prediction of the salient properties of the products and help to acquire an insight into the mechanisms involved in processing at this level—to understand the fundamentals of processing at micrometer, nanometre, or molecular scales. Processes can be classified according to their precision. This is especially important in the case of the material removal processes where the search for ever greater accuracy is continuous. According to Taniguchi, three categories can be distinguished: normal, precision, and ultra-precision processes [4]. The borders between each pair of categories are defined as a function of process precision relative to that of other processes, for example, what was considered to be ultraprecise in the 1980s is today only precise, due to advances in technology. There is, therefore, no uniform definition of ultra-precision processes, other than that they are the most extremely accurate processes of their epoch.

Micromachining is frequently used for describing ultra-precision processes aiming toward the production of micro parts and miniaturized components. Miniaturization machining operations have evolved into one of the key technologies in the field of micro and nano processes, yet a commonly accepted definition of micromachining is lacking. Masuzawa states that “micro” literally represents the range between 1 and 999 μm [5]. However, since small components introduce difficulties during their manufacturing compared to macro components, “micro” should incorporate the meaning that something is too small to be machined easily. Furthermore, taking into account the epoch, machining method, type of product or material the range 1–500 μm was finally adopted to delineate the upper and lower limits of micromachining. However, it is very common to shift this range to 0.1–100 μm and include, as well as the size of a component, its accuracy or surface texture [3, 6] Moving downwards from 0.1 μm (=100 nm) the region defining nanotechnology is entered. A broadly accepted definition is that nanotechnology pertains to the processing of materials in which structure of a dimension of less than 100 nm is essential to obtain the required functional performance [7]. Similarly to micromachining, nanotechnology’s regime extends from 0.1 to 100 nm and includes the ability to observe and measure features at this scale. Theoretically, the limit of nanotechnology is the processing of atoms, the diameter of an atom being about 1 Å. Figure 1 illustrates some components produced by micro- and nano processes in order to apprehend the size of such components.

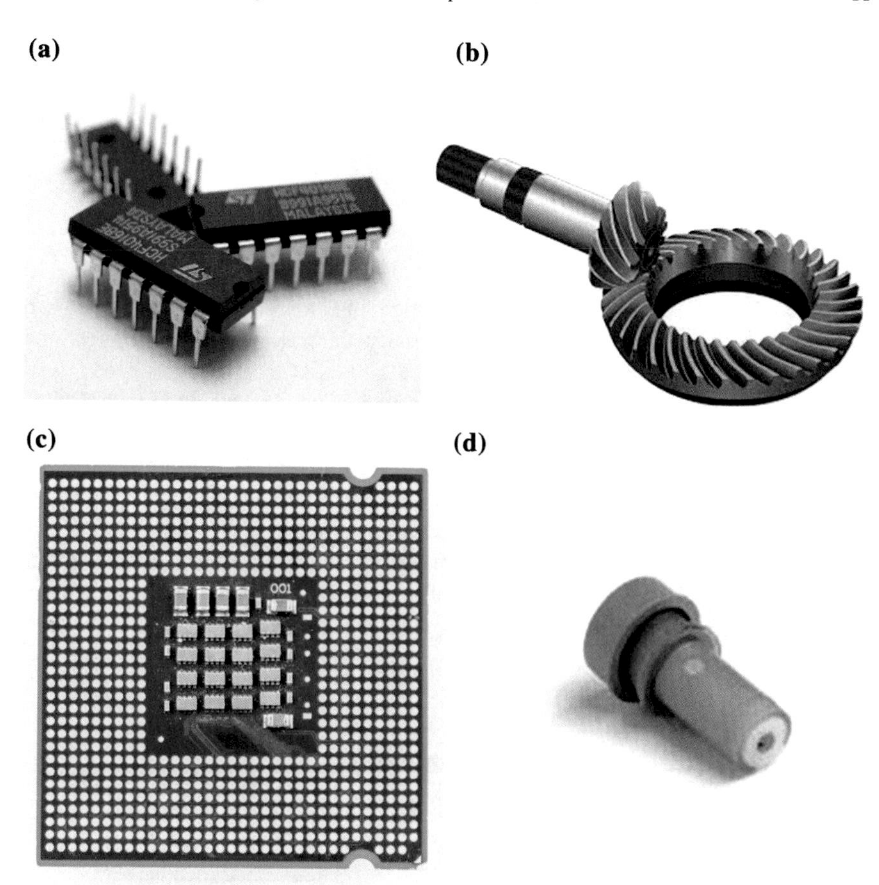

Fig. 1 Components produced by micro and nano processes **a** transistor on IC (By Kimmo Palosaari (OpenPhoto.net) [Public domain], https://commons.wikimedia.org/wiki/File: Three_IC_circuit_chips. JPG">via Wikimedia Commons. Link: https://upload.wikimedia.org/wikipedia/commons/8/80/Three_IC_circuit_chips.JPG), **b** Micro gears (https://commons.wikimedia.org/wiki/File:Gear-kegelzahnrad.svg#/media/File:Gear-kegelzahnrad.svg), **c** Microprocessor (Attribution: Raimond Spekking/via Wikimedia Commons https://upload.wikimedia.org/wikipedia/commons/3/37/Intel_microprocessor_Pentium_4_HT_651_3.4_GHz_-_SL9KE-3368.jpg), **d** micro pump (By Sensile Medical AG (www.sensile-medical.com) [CC BY-SA 3.0 (https://creativecommons.org/licenses/by-sa/3.0)], via Wikimedia Commons https://upload.wikimedia.org/wikipedia/commons/c/cb/Micro_Pump.png)

The aim of this chapter is to describe some micro and nano processing techniques, present their applications in contemporary industry and make speculations for the future. It is not possible to discuss all the techniques available today within this chapter but some of the most applied and the most promising for the future are selected; their main characteristics, capabilities, and applications are outlined. Most of the techniques presented are material removal processes. Material removal is achieved by various methods according to the principle the process is based on. In the following paragraphs, the processes presented are categorized according to these principles. Their advantages and disadvantages and the precision they can attain are discussed and their capabilities are compared.

2 Micro and Nano Processes Classification

Micro and Nano machining can be classified into a lot of ways as there is no specific way to classify, though some classify them according to conventional/traditional and nontraditional way some on the basis of the processes used [8–10]. Micromachining and nanotechnology can be achieved by means of various methods and techniques. However, in all the cases examined the trend toward smaller dimensions, higher accuracy, and production of components that are highly functional is common. The development of new technologies follows mainly two directions: the downscaling of manufacturing processes that have an existing background as conventional ones and are already widely used in industry, and the development of new ones that are suitable only for this kind of manufacturing. A brief classification of micromachining is shown in the Fig. 2.

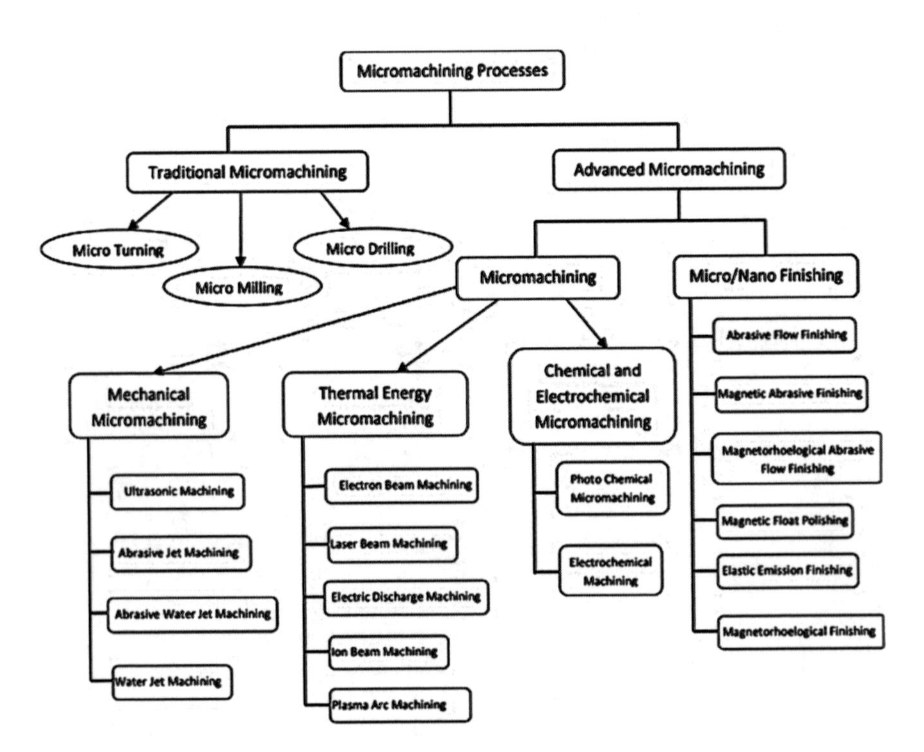

Fig. 2 Classification of micromachining processes

It is usual, and this classification will be followed hereafter, to distinguish between lithography and non-lithography-based micro and nano processes. This is due to the widespread of lithographic processes, especially photolithography, for the manufacture of ICs [11]. Although photolithography has served for many years as the exclusive process used for silicon chips it seems that it has reached its limits and that is why new lithographic processes are being introduced [12, 13]. Some of these processes, often called Next-Generation Lithography (NGL), are presented in what follows and their advantages and disadvantages are discussed. X-ray lithography, Lithography, Galvanoformung and Abformung (LIGA), electron beam lithography, and 3D lithography among others will be discussed.

The non-lithography-based processes include mainly machining operations that use mechanical, thermal, electrothermal, or electrochemical energy in order to achieve material removal. In this category, conventional and nonconventional machining processes are included, suitably altered to perform in the micro world. Some of the processes examined are micro cutting, micro-Electrical Discharge Machining (micro-EDM), laser processes, and electrochemical machining [14, 15].

Our aim is not to describe all micro and nano processes thoroughly and in detail; for that, an adequate number of references to the relevant literature is provided. The aim is rather to describe some of them, depict some of their important characteristics and emphasize the differences that make these processes unique and promising. In the case of lithography-based processes, photolithography is the main subject. NGL is also analyzed and their advantages or disadvantages in comparison to photolithography are pointed out. In the case of non-lithography-based processes, a variety of them is described and compared. Furthermore, owing to the fact that they are also realized as macro as well as micro processes, the differences between the two types are elaborated. This analysis points out the difficulties arising from transferring technology between the macro- and micro worlds and leads to the fact that sometimes a feature that is considered as an advantage in one world may be a disadvantage in the other and vice versa. Some important applications are described so that the reader can have an idea about how these technologies are applied in everyday life.

3 Lithography-Based Processes

Lithography is a combination of two Greek words, namely “lithos” (stone) and “graphein” (write), and refers to a kind of art invented in the late 1700s involving the transfer of an original image or pattern carved on a stone onto paper. This kind of art has been the inspiration for a technology, the most widely used form of which is photolithography (from another Greek word meaning light), used in the IC industry for the production of chips [16].

4 Photolithography

In Fig. 3, the basic steps followed during photolithography can be seen. In this example, an oxidized Si wafer is used and a simple pattern must be transferred to it. In Fig. 3a the oxidized wafer is coated with a photoresist layer (resist), a polymer sensitive to ultraviolet (UV) light approximately 1 μm thick. Then, a mask is placed on the resist, Fig. 3b. The mask (also referred to as a photomask) is the stencil used to repeatedly generate a desired pattern on the resist coating to be finally transferred to the oxide. It is an optically flat glass, some parts of which are covered with a metal layer, usually chromium, forming the pattern. The glass parts of the mask are transparent to UV while the chromium plated ones absorb it. The resist is exposed to UV, which passes through the mask, and depending on its type, so-called negative or positive, the resist is altered. In negative resists, UV hardens the areas exposed to it, while in positives the opposite happens. In Fig. 3c, an example of a negative resist is given, where the negative image of the mask is transferred onto it.

After exposure, the wafer is rinsed in a developing solution or sprayed with the developer, which removes the unexposed areas of the resist and leaves a pattern of bare and resist-coated oxide on the wafer surface, Fig. 3d. In the next step after development, the wafer is placed in a solution of hydrofluoric acid (HF) or acid buffered with ammonium fluoride ($HF + NH_4F$), meant to attack the oxide but not the resist nor the underlying silicon; the resist protects the oxide areas it covers, Fig. 3e. Once the exposed oxide has been etched away, the remaining resist can be stripped off with a strong acid, such as sulphuric acid (H_2SO_4), attacking the resist but neither the oxide nor the silicon, Fig. 3f. The oxidized Si wafer with the etched cavity in the oxide, Fig. 3g, serves as a final product that has been appropriately formed or can be further processed. For example, the cavity can be filled by depositing a desired material, or a new resist layer can be added and a new feature with a new mask can be created, repeating the steps described in Fig. 3a–f.

Negative and positive resists have characteristics that make them suitable for certain applications in the IC industry. These characteristics have to do with their ability to adhere to Si, the minimum feature size that can be produced, their available compositions and the developers used, thermal stability and other properties, and cost [6, 17]. Moreover, there are also permanent resists that are not removed during the process, in contrast to the resist in the example. As far as the position of the mask relative to the resist is concerned, there are a few possibilities: the mask can be in actual contact with or merely in proximity to the resist. In the first case, the hard masks, as they are called, wear easily, and cannot be used many times. In the second case, soft masks are placed approximately 10–20 μm above the resist. In both cases, described as shadow printing, the pattern is transferred as a 1:1 image. A more reliable method is projection printing, where the mask is focused by a high-resolution lens system onto the resist. That way the mask is not subjected to wear while reduction of the pattern by a factor 1:5 or 1:10 is possible. Additionally, step-and-scan systems are used, where the mask or the wafer are moving and after the image of one chip is printed at a certain location of the wafer, the system steps to the next location where the next chip is printed, and so on. In this way, hundreds of ICs are printed on one wafer and productivity is enormously increased.

Small sizes and continuous further shrinkage in the IC industry are indeed essential for its future. Chip structures have become very compact and a large number of transistors packed closely together has allowed computer speeds to increase since the electrical signals have less distance to travel between two transistors. Small microchip sizes have resulted in smaller, less power-consuming and yet more powerful computer systems. In order to keep up with Moore's "law" further shrinkage of individual transistor dimensions must be achieved, which requires further refinement of the manufacturing. New techniques are being introduced that aim at the improvement of photolithography's resolution. These techniques, collectively called resolution enhancement technologies (RET), focus on the improvement of resists and masks. Resist improvement includes chemically amplified resists, such as SU-8, and resists that are sensitive to shorter wavelengths, while mask improvement focuses on techniques that can reduce the bothersome diffraction of light, using phase-shifting masks (PSM), and grey-tone masks (GTM) that notably increase the efficiency of optical elements. Nevertheless, the smallest feature that can be manufactured, even with these improvements, is about half the wavelength of the radiation used. The usual excimer lasers used today as UV sources have wavelengths of a few hundred nm resulting in features with dimensions around 100 nm. The search for shorter wavelengths and improvements in photolithography for the manufacturing of microstructures has led to the development of new lithographic techniques known as NGL technology, which is discussed in the following paragraphs.

5 Next-Generation Lithography (NGL)

Continuous improvement of traditional photolithography has delayed the development of wholly new methods for the fabrication of ICs. Equipment cost and technical reasons have been obstacles in the research and development of NGL, and their industrial adoption [18]. The methods discussed below are probable candidates for the production of chips in the future.

A plausible extension of UV photolithography is Extreme UV Lithography (EUVL), which uses laser-produced plasmas, or synchrotrons, to generate wavelengths of 10–14 nm. The very short wavelength is an advantage but unfortunately, EUV is absorbed by almost every material and the process must take place in a vacuum [19]. Furthermore, all the usual optical accessories are reflective and not refractive at this wavelength and new resists must be developed that are sensitive to this wavelength. There are also some technical problems related to EUV sources and a lot of research is currently focused on that topic [20–22].

Another promising technique is X-ray lithography, where instead of UV light X-rays are used. The X-ray wavelength is about 10 Å and diffraction effects are negligible. This method uses no optical equipment, which seems to be an

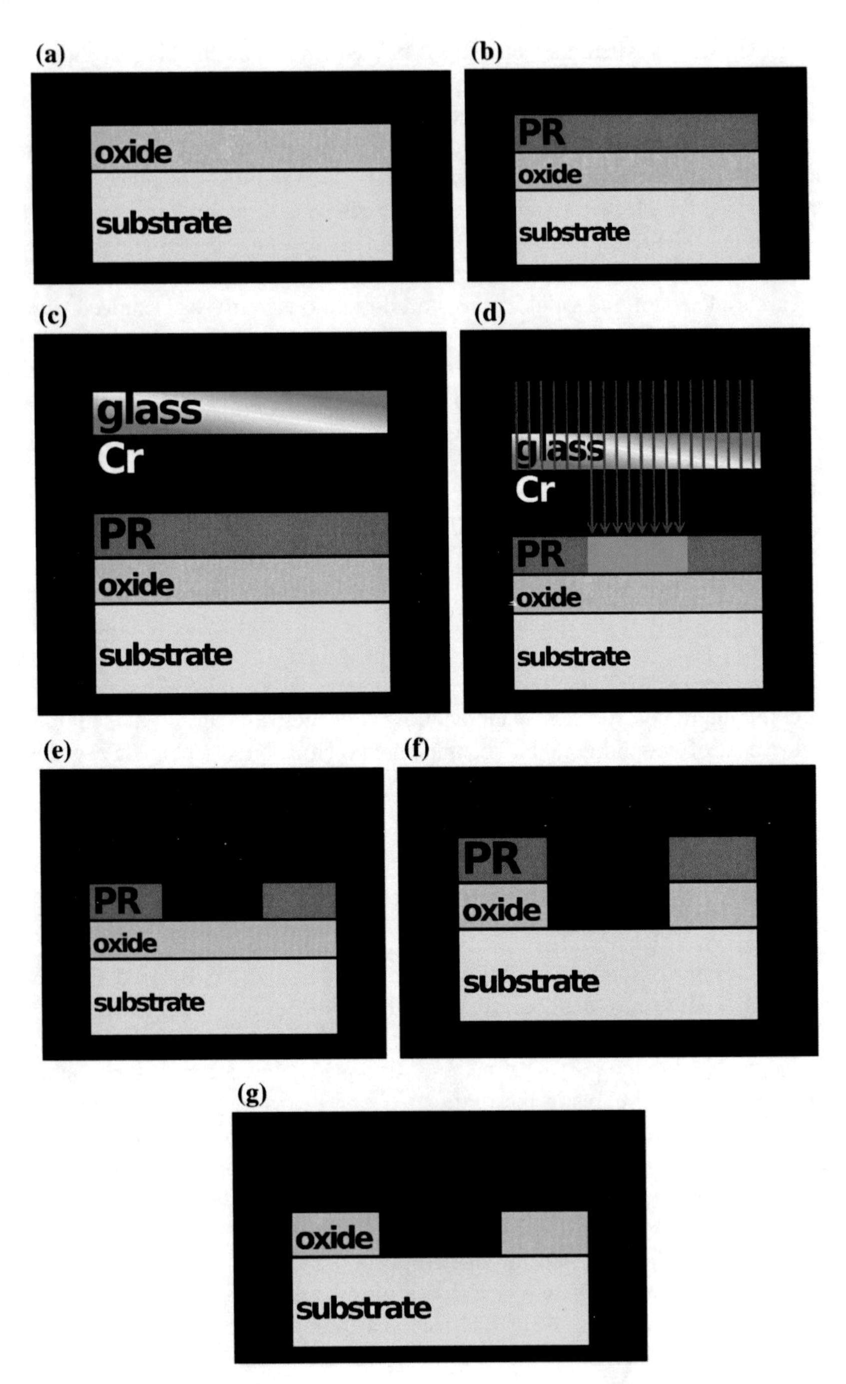
(a)
oxide
substrate
(b)
PR
oxide
substrate
(c)
glass
Cr
PR
oxide
substrate
(d)
glass
Cr
PR
oxide
substrate
(e)
PR
oxide
substrate
(f)
PR
oxide
substrate
(g)
oxide
substrate

◀**Fig. 3** Photolithography process (By Photolithography_etching_process.svg: Cmgleederivative work: Cepheiden (Photolithography_etching_process.svg) [CC BY-SA 3.0 (https://creativecommons.org/licenses/by-sa/3.0) or GFDL (http://www.gnu.org/copyleft/fdl.html)], via Wikimedia Commons https://upload.wikimedia.org/wikipedia/commons/f/fa/Photolithography_etching_process_%28DE%29.svg), **a** prepare wafer, **b** apply photoresist, **c** align photomask, **d** expose to UV light, **e** develop and remove photoresist exposed to UV light, **f** etch exposed oxide, **g** remove remaining photoresist

advantage, but is also a disadvantage because only 1:1 shadow printing can be performed. Thus, the mask must have the same dimensions and dimensional tolerance as the product, making its manufacturing a real challenge for micro- and nano processes [23]. The most important technical problem is, however, the X-ray source that should be used. One possibility would be to use the radiation emitted by electrons as they circulate in a synchrotron storage ring, but the cost for such an installation is prohibitive [24]. Some industries have developed their own X-ray lithography systems and use it mainly for the production of electronics. Indeed, a method for producing 3D structures using X-ray lithography has been reported. Conventionally, during the X-ray exposure, the mask and the substrate are mounted perpendicular to the beam, resulting in vertical profiles. When the mask and the substrate are tilted with respect to the beam, inclined profiles may be produced.

LIGA is a process based on X-ray lithography and involves some additional steps. In its original form, in the first step of the process X-rays are used to expose a thick layer of resist to a depth up to 1000 μm, with a lateral resolution of better than 1 μm. After exposure, a resist mold is produced that is the replica of the mask pattern. Then, depending on the material and number of parts required for the final product, different fabrication routes can be chosen: the polymer can be used as it is, as in X-ray lithography, or it can be subjected to electroforming and molding techniques [25]. In the application of electroforming techniques, the polymeric mold is filled with metal or ceramic. The resist is removed and metallic or ceramic micro parts are produced. Alternatively, by the same method, a metallic mold is produced that can be used several times to mold replicates from other materials, primarily plastics. Several types of molding processes have been used and the resulting plastic part may again serve as a mold, like the original resist structure, for fast and cheap mass production since there is no need for a new X-ray exposure. LIGA is characterized by tight tolerances and high precision, and 3D structures can also be constructed by tilting angles, as described above. Sometimes, deep X-ray lithography is used as a first step, or other sources of radiation such as conventional photolithography or UV light from low-cost excimer lasers. The applications of the method are many and interesting and involve the manufacturing of MEMS, micro motors, sensors, nozzles, molds for micro-fibers, optics, and high precision tools [26–28].

A process that allows the fabrication of lines much less than 100 nm thick is Electron Beam Lithography (EBL), which uses high-energy electrons to expose electron-sensitive resists. This method, like X-ray lithography, does not practically limit the obtainable feature resolution due to diffraction and furthermore, there is no need for a physical mask, but a software mask is used instead. The pattern is stored in a computer and the beam is directed to the appropriate area enabling the pattern to be written sequentially, point by point, over the whole wafer. This makes the process rather slow, even though arrays of electron beams can be used [29]. Other disadvantages of the method are that electrons readily scatter in solids, limiting resolution to dimensions greater than 10 nm and that the process has to be performed in a vacuum. All these and the high cost of the equipment have limited the application of EBL to specialized applications, including mask-making for other lithographic techniques.

A technique known as SCALPEL (Scattering with Angular Limitation Projection Electron Beam Lithography) has been developed by Lucent Technologies. It is a projection electron beam technique employing a step-and-scan system (Fig. 4). A membrane made from a low atomic number compound, such as silicon nitride, is used as a mask, with an overlaid high atomic number material (typically tungsten) used to make the pattern. High-energy (100 keV) electrons uniformly illuminate the mask and electrons passing through the pattern are scattered while those passing through the membrane suffer very little scattering. A projection system with two lenses performs a 4:1 reduction of the pattern. An aperture in the back focal plane of the second lens stops the strongly scattered portions of the beam, producing a high contrast image at the wafer plane. The small images at the wafer plane are stitched together by suitably moving the mask and the wafer, which are servo-controlled using a laser interferometer system. A system named SCALPEL-Proof of Concept was created in 1996 for testing the method and features of 80 nm were produced [30].

A related kind of processing is Ion Beam Lithography (IBL), which is a category of lithography-based processes with great future potential. One kind of IBL is Focused Ion Beam (FIB) machining, where electrons are replaced by high-energy ions of a volatile metal (typically gallium) [31]. Beam diameters of less than 50 nm are achieved and as in the case of EBL, no mask is required. Other variants are Deep Ion Beam Lithography (DIBL), which employs high-energy protons, and Ion Projection Lithography (IPL) where protons are generated by a radio-frequency-driven filament [6, 17, 32]. IBL has certain advantages over EBL; it has a better resolution because ion scattering is negligible in the resist and the resist sensitivity is much higher. IBL spot size is the smallest possible among UV, X-ray and electron beam spots, the FIB spot size being about 8 nm.

Traditional lithography techniques result in projected 2D structures rather than 3D shapes. With the exception of some NGL techniques mentioned above, lithography cannot produce curved surfaces, which are sometimes needed in micro- and nano applications. For facing this problem 3D lithography method have been developed, such as holographic lithography (HL) and stereolithography (SL). In HL the mask is replaced by a holographical constructed one and a hologram of the pattern is made on the resist layer. In SL, in a procedure similar to rapid prototyping, light exposure solidifies liquid resin into the desired 3D shape. Research is underway to improve these processes and their applications, especially in the fields of micro part construction, biomedical engineering, actuators, and tools [6].

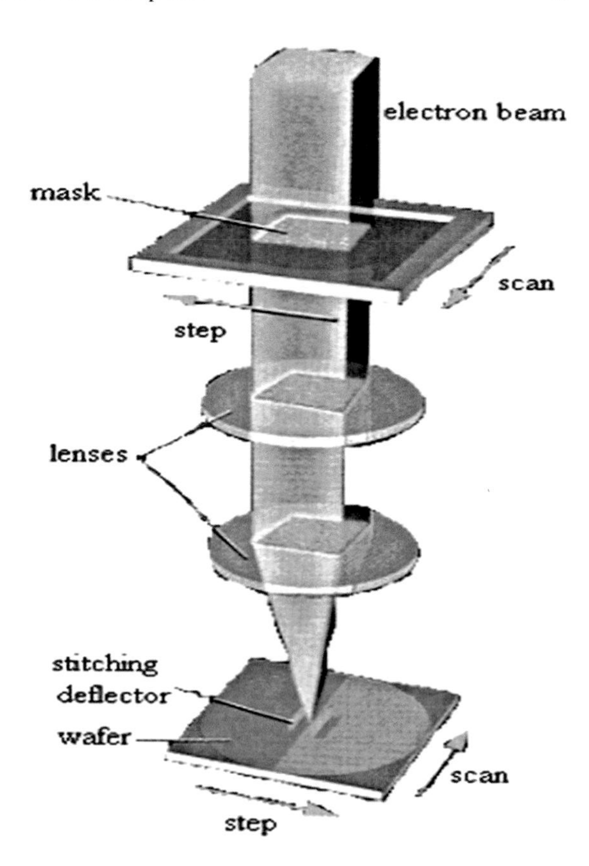

Fig. 4 SCALPEL technique illustration

6 Non-lithography-Based Processes

The processes included in this section are conventional and nonconventional machining processes, which are being used in traditional manufacturing as well as in the nano realm. They are at the forefront of industrial integration and their applications have reached a high level of production maturity. Some of them have the prefix "micro" in their names to declare that they are processes following the same principles as the original macroscopic ones, but particularly designed as microprocesses. A wide variety of processes could be included here, but to avoid undue length some characteristic ones only will be described. It should be noted that unlike lithography-based processes they are not used in the IC industry, but this certainly does not mean that they are not suitable for micro and nano processing. On the contrary, they have advantages related to the fact that they can produce 3D structures for a variety of materials, including metals and hard and brittle substances with fine tolerances and high accuracy. The processes described will be subdivided into categories according to the kind of energy they use for machining, namely mechanical, electrothermal, thermal, and electrochemical.

7 Mechanical Processes

Mechanical processes are probably the most popular among the micro-processes in current use. They involve mechanical interaction between a sharp tool and the workpiece causing the removal of unwanted material in the form of a chip [33, 34]. Conventional machining operations such as turning, milling, grinding, and drilling belong to this subdivision [35]. Advances in the subsystems involved in macroscopic machining such as positioning, automation, numerical control, metrology, and tools have made it possible to apply them in microfabrication [36, 37]. In particular, the "micro" versions of the aforementioned processes are used for the production of miniaturized parts, for drilling microholes, for shaping microgrooves and to achieve mirror-like super finished surfaces.

Micro turning makes use of very small, single crystal diamond tools with edge radii less than 1 μm. They are used for very small depths of cut, in machines that can reproduce movements with 5 nm positioning accuracy in different directions, and achieve roughness down to 50 nm. Actually, this is problematic with ferrous materials, but CBN tools are developed for this particular case [38, 39].

Grinding is a process traditionally used for finishing operations, thus it is considered to be the most precise mechanical process. Nevertheless, in micromachining, grinding is not necessarily superior to other processes, but smooth surfaces of less than 10 nm peak to valley have been reported with this process [3, 40]. Such smoothness can be realized either by grinding wheels with coarse abrasives mounted on high precision machine tools and with the use of accurate dressing or by grinding wheels with fine abrasives (particles having sizes of about 10–20 nm). An abrasive process, referred to as nano grinding, has been reported as yielding an average surface roughness of 1.14 nm for Al2O3-TiC and 0.79 nm for SiC, measured by atomic force microscopy [41].

The advantages of the micromachining processes presented so far in this section are the very extensive knowledge that has been accumulated with their macroscopic precursors, the fact that they are relatively low-cost processes, and that they allow the manufacture of quite complex 3D structures. A wide variety of materials can be machined, including metals, plastics, and their composites, since the electrical properties of the workpiece do not influence the process, in contrast to the processes to be discussed in the following paragraphs. A typical application example is the drilling of microholes in laminated printed circuit boards. A disadvantage is the relatively high machining force, which limits machining accuracy due to deflexions of the tool and the workpiece [42].

Micro-ultrasonic machining (micro-USM) is another mechanical micro process having its origins in a traditional macroscopic process. It employs a tool and a mixture of a fluid (water or oil) with abrasive particles. The tool is vibrated at an ultrasonic frequency and drives the abrasive to create accurately shaped cavities on the surface of the workpiece [43]. The shape and size of the cavities depend on those of the tool. In micro-USM, micro tools and fine abrasives are used, with which ± 10 μm tolerance can be achieved. Since the tool does not exert any pressure on the workpiece the method is suitable for machinings hard and brittle materials such as aluminum oxides, silicon and glass. Furthermore, the method is non-thermal and as a result, it produces stress-free surfaces without defects attributed to heat-affected zones. Micro-USM has low operating costs, the required operator skill level is modest and the production rates are relatively high [44]. Technical problems are connected with the accuracy of the tool holding and the response of the equipment itself to the ultrasonic vibration. In order to overcome the first problem, on-machine tool preparation has been introduced, in which the tool, before its preparation, is soldered to the machine head and then machined to the desired dimensions and shape. An approach to overcome the second problem is to apply the vibration to the workpiece instead of the tool [45].

Abrasive Jet Machining (AJM) is an operation where grains, with sizes less than 100 μm, impinge with high velocity on the workpiece surface whose material is removed by the impacts. This method is used for making accurate shallow holes in electronic components, and, with the use of masks, patterns on semiconductors [46]. The process is fast and the equipment inexpensive. Minimum feature sizes that can be made with this process are less than 50 μm.

FIB was described earlier as an ion beam lithography method. Recently, it has been suggested that it is suitable for producing microcomponents without the requirements of lithography. Sometimes FIB is considered as a thermomechanical process but it is more widely accepted as a solely mechanical technique in which the tool is replaced by a stream of energetic ions. The advantage of the method is that it can process very hard materials into peculiar shapes, but it is expensive and slow [47]. A variation of FIB is Fast Atom Beam (FAB) machining where ionized atoms are accelerated to high speeds, neutralized and used as a mechanical cutting tool.

8 Electro-Thermal and Thermal Processes

In electrothermal and thermal processes thermal energy is provided by a heat source that is used to melt or vaporize the material to be removed. Machining forces are very small, and that permits the use of small and thin tools. The mechanical properties of the workpiece do not influence the machining process but thermal properties and in some cases electrical ones are important. In these processes, it must be taken into consideration that because the tool is not in contact with the workpiece, uncertainties are introduced in specifying workpiece dimensions. Furthermore, heat-affected zones (HAZ) may appear in the workpiece. Heat effects include resolidifying debris on the surface and metallurgical transformations undergone by the layers just below the surface, which alter the properties of the material as a whole and may cause problems.

Electrical Discharge Machining (EDM) is the oldest and most-used operation of this kind. Micro-EDM is already widely used for micromachining [48]. The workpiece (anode) and the tool (cathode) are submerged in a dielectric fluid and subjected to a high voltage. When the electrodes are separated by a small gap (whose dimensions can be precisely calculated) a pulsed discharge occurs. Sparks are generated and the material is removed through local melting and evaporation. Both electrodes are worn away but the tool wear ratio varies depending on the tool material and can reach values up to 70:1 for carbon electrodes. Because there is no contact between the electrodes, machining forces are negligible and the hardness of the workpiece is not critical, making the process eligible for machining conductive, hard and brittle materials. Machining of non-conductive ceramics is possible under certain conditions, but the method is still under development. Machining accuracy and repeatability of the process are good and the structures are burr-free, but the material removal rate is slow. By suitably adjusting process conditions surface roughness of 0.1 μm can be obtained, but the material removal rate is slower than if rougher surfaces are achieved.

Current micro-EDM technology can be categorized into four types [49]. Sinker micro-EDM is performed as described above and is used for the fabrication of mirror image replicates of the tool on the workpiece. In micro-wire EDM a wire diameter as small as 10 μm is used to cut through a conductive workpiece. This technique is used for producing microrods by rotating a cylindrical workpiece against the cutting wire. Micro-EDM milling is another variant, in which microelectrodes 5–10 μm diameter are employed to produce cavities in a similar way to that of milling. Finally, micro-EDM drilling is a process where rotating microelectrodes are used as drills to create microholes in the workpiece. The holes can be 5–300 μm in diameter with depths up to 5 times the diameter and can have a precision of ±0.5 μm circularity. EDM and micro-EDM are well-established processes used for the production of molds, extrusion dies, fuel injection nozzles, turbine blades, microholes, grooves, boreholes, complex 3D structures and convex shapes, and a lot of research is directed toward the improvement of the process and its applications [50–52].

An important thermal technique is Electron Beam Machining (EBM). In this process, instead of electrical sparks high-velocity electrons, traveling at about three quarters the speed of light, are used. The electrons are focused on an area on the surface of the workpiece and on the impact their kinetic energy is converted into thermal, causing the workpiece material to melt and vaporize. The diameter of the beam is 10–200 μm, producing holes below 0.1 mm in diameter, with tolerances about 10% of the diameter and with a diameter-to-depth ratio of 1:10. With multiple pulses, this ratio can be extended to 1:100. The process is performed in a vacuum to prevent scattering of electrons by gas molecules of the atmosphere. Once the vacuum is established the process is extremely fast and is used for difficult-to-machine materials. A serious disadvantage of the method is its cost, which is high compared to other micro- and nano processes [53].

Laser Beam Machining (LBM) is perhaps the most prominent process type presented in this section [54]. LBM was introduced in the industry for macroscopic cutting and welding of metals but lately is being used in micromachining as well. The importance of lasers in photolithography as UV sources was already discussed; here their application in ultra-precision material removal will be analyzed. A thin laser beam is focused to a small spot on the surface of the workpiece and material is removed by ablation. The process is controlled by several parameters, including the light wavelength, others being spot size, beam intensity, and depth of focus of the beam.

Laser wavelength λ is important because the minimum size of a feature that can be fabricated with light is about λ/2. The spot size, i.e., the minimum diameter of the focused laser beam, also depends on the wavelength. It is generally accepted that for wavelengths shorter than 200 nm direct photochemical bond breaking plays an important role in ablation. Laser types currently used in micromachining are solid-state lasers, such as Nd-YAG and Ti-sapphire lasers with wavelengths of 1.064 μm and 775 nm, respectively, and gas lasers, such as CO2 ($\lambda = 10.6$ μm) and excimer lasers. Excimer (a contraction of EXCIted diMER) lasers are pulsed lasers driven by a fast electrical discharge in a high-pressure mixture of a rare and a halogen gas. Available wavelengths are 353 nm (XeF), 308 nm (XeCl), 248 nm (KrF), 193 nm (AF), 157 nm (F2) and several years of research and continuous improvement has enabled them to be successfully applied in ultra-precision processes, especially for the fabrication of very complex 3D structures [55, 56]. These lasers are more advantageous than other lasers because the wavelengths are more compatible with micromachining and they produce less thermal damage. It should be noted that typical powers for these lasers range from 3 to 50 MW (with pulses lasting a few tens of ns), with which almost all metals can be machined.

Based on pulse length three operation modes can be specified: long, short and ultrashort. In long mode the pulse duration is greater than 0.25 ns and is of limited use in micromachining mainly due to the generation of deleterious HAZ: heat diffuses to the surrounding material causing undesired results. Short pulses last no longer than 10 ps, and ultrashort pulses are a million times shorter than nanosecond pulses and include the so-called femtosecond lasers. Femtosecond lasers have peak powers of 5–10 GW, powers that no material can withstand. This means that they can be utilized for the processing of metals, ceramics, glass, polymers, semiconductors, including very hard materials, even diamond, and materials with high melting points, such as molybdenum. Another advantage of machining with femtosecond lasers is that there is no damage to the material because the duration of the laser pulse is shorter than the heat diffusion rate. This increases the efficiency of the process and is beneficial for the obtainable precision of the machining [57].

9 Electrochemical Processes

In this subsection, electrochemical energy is involved in the material removal or forming operations. Additive techniques and the main processes for the production of porous Si are also included here. The most popular processes are electrochemical machining (ECM) and electrochemical grinding (ECG). ECM involves a cathode electrode and an anodic workpiece which are separated by a highly conductive electrolyte. When current is applied material removal is accomplished [58]. Like in EDM and USM the shape of the tool electrode defines the shape obtained on the workpiece. ECM is suitable for micromachining because removal is achieved virtually atom by atom. In ECG, a conductive abrasive wheel is used as cathode and material is removed both electrochemically and by abrasive action.

Machining with ECM results in very smooth surfaces. Metals, regardless of their physical properties, especially their hardness, can be machined as in EDM. Compared to the latter, ECM has the advantage that no HAZ are created, and there is almost no tool wear. Other advantages are the low running costs, the by-production of very little scrap, and the ease of automation of the operation, while machining time is comparable to that of EDM. Furthermore, no residual stresses form and it is a burr-free process, also used for deburring machined components. Disadvantages of the method are the initial cost of the equipment and the fact that dissolution of the material occurs over larger areas than those facing the electrode. Electrochemical processes, despite their advantages, are not used in the IC industry, because in this particular field the so-called "dry processes" are preferred. Generally speaking, non-lithography-based processes are of limited application in the IC industry due to the reasons explained in the following paragraphs, but extremely useful in other areas of application.

10 Nanofabrication Methods

This section is dedicated to methods that are used for manipulating matter atom by atom in order to fabricate stable structures at the nanometre scale. This "bottom-up" approach, where instead of subtracting matter, atoms are placed in prescribed positions to form a complicated, functional structure, is somewhat different from the methods described so far. Some scientists have even envisaged the fabrication of nanomechanical components consisting of atoms, like the planetary gear. Molecular nanotechnology, as this approach is called, became known through the pioneering work of K. E. Drexler, who has worked on establishing its basic concepts and describing manufacturing systems able to produce components and devices of molecular size [59].

Two instruments, the scanning tunneling microscope (STM) and the atomic force microscope (AFM), are used to fabricate surface structures with dimensions from less than 100 nm to atomic dimensions. The STM was developed in the laboratories of IBM in Zürich and originally it was built for studying the topography of material surfaces at the atomic level. Its inventors, who won the Nobel Prize in 1986 for their work, were able not only to measure and image surfaces at the molecular scale, but also to manipulate atoms and place them in designated positions [60] With this instrument 35 xenon atoms could be arranged on a nickel substrate, spelling out 'IBM'. AFM is a related instrument capable of producing 3D images of any surface, whereas the STM is only useful for conducting surfaces. Since their invention, about 20 years ago now, these instruments have added considerably to nanotechnology research but also have numerous applications in metrology, electronics, data storage and many other fields [61].

11 Future Scope

Lithographic processes are already deeply embedded in manufacturing in many fields of science and engineering, and especially in the IC industry. The other processes, although little used in the IC field, are preferred for the commercial production of, e.g., computer hard disks, ink-jet printer nozzles, mirrors, lenses, compact disk reader heads, photocopier drums, telecommunications devices, MEMS and many others, exploiting the very small sizes and excellent surface finishes that can be achieved and the wide selection of materials and geometrical characteristics that can be handled. The preference for lithography techniques for ICs is dictated by both technical and economic reasons since ICs require the manufacturing of very small features and batch fabrication techniques. On the other hand, devices such as MEMS, which are made for very specific purposes, are produced in low volumes and typically use non-IC materials. For such devices, for which low cost is not as important as for, e.g., computer chips, other processes may be useful.

There has been a significant rise in the number of individual electronic components in a microprocessor which has increased from 20,000 transistors in 1980 to hundreds of millions on the latest silicon chips, with parallel increases in computer power and memory. Computer speeds have increased to an incredible number all because of a large number of transistors used and chip structures becoming more compact. In a process similar to the operation of the phonograph, the scanning tip of an AFM is heated and then makes contact with a disk. The tip, whose trace is about 40 Å, "burns" and writes the data on the plastic material of the disk. Data reading is performed by bringing an AFM tip into contact with an already written disk. With this method, readable data storage with a density of 400–500 Gb/in2 can be achieved, which is quite an accomplishment if one considers that conventional magnetic data storage devices have densities of 20–50 Gb/in2. Such AFM-based data storage devices are now being studied at IBM, with especial regard to control of contamination of the material during its production and processing. Their small size will enable them to be used in watches, cellular telephones, laptop computers, etc., whilst their high-density data storage capability will lead to terabit data storage systems on 2.5 in. hard disks.

The modern automotive industry widely incorporates electronics into cars. These devices are mainly miniaturized sensors. Some of the current devices in automotive applications are pressure sensors, gyroscopes, fuel, and air flow control systems, accelerometers and microactuators which are quite common in automobile these days. They result in enhanced car performance and safety and also reduced component cost. Common examples are the airbag accelerometers which helps and protects the driver in sever crashes from getting injured, intelligent sensors that efficiently controls engine and anti-skid and roll-over systems. The shrinkage of electronic systems (as well as the associated engineering components) has led to significant reduction of the size and weight of car systems. A typical example is the ABS for braking, whose weight has been reduced from 6.2 to 1.8 kg, according to Bosch, from 1989 to 2001. Note that several car parts must now be ultra-precision machined in order to be functional.

In space applications, the trend toward miniaturization is probably greater than in any other field. Demand for smaller satellites that orbit the earth has increased over the past few years, due to growth in communications as the internet, mobile phones, television stations, etc., all require satellites. Industries and enterprises are putting a lot of effort to make them smaller because they are then easier to put into and maintain in orbit, and cause minimal pollution when going out of order. Space agencies are developing small space probes able to travel to other planets.

Medical applications are very important for the improvement of the quality of life. They include prosthetic implants and surgical tools which are ultra-precision finished, and diagnostic devices small enough to enter the body of a patient. Another area is the development of nanostructured drug delivery particles, with selectively reactive molecular coatings that will act in specific places inside the human body.

12 Conclusions

In this chapter, the micro and nano processes used today and considered to have the potential to play a major role in tomorrow's microfabrication have been presented. Both lithography and non-lithography-based processes have been considered and their main characteristics have been described and compared. Furthermore, some nano-fabrication methods have been discussed. Finally, some current and potential uses of these processes have been analyzed to show their importance in today's industry and life. Even though the impact of nanotechnology is already great, expectations for tomorrow are even greater, as it will make health treatment, communication, transportation, data storage, automobile, and aviation sector and many other scientific and industrial applications faster, safer and much cheaper. The advantages of such items and applications in numerous technological areas will be critical and the fields to which they will be applied will lead to a new era. International interest is greatly increasing and leads to a concomitantly growing research activity, in the very vanguard of modern science and technology.

References

1. X. Luo, K. Cheng, D. Webb, F. Wardle, Design of ultraprecision machine tools with applications to manufacture of miniature and micro components. J. Mater. Process. Technol. **167**(2–3), 515–528 (2005). https://doi.org/10.1016/j.jmatprotec.2005.05.050
2. E. Uhlmann, M. Röhner, M. Langmack, T.-M. Schimmelpfennig, Micromanuf. Eng. Technol. (2015). https://doi.org/10.1016/B978-0-323-31149-6.00004-9
3. L. Alting, F. Kimura, H.N. Hansen, G. Bissacco, Micro engineering. CIRP Ann. Manuf. Technol. **52**(2), 635–657 (2003). https://doi.org/10.1016/S0007-8506(07)60208-X

4. N. Taniguchi, Current status in, and future trends of, ultra precision machining and ultrafine materials processing. CIRP Ann. Manuf. Technol. **32**(2), 573–582 (1983). https://doi.org/10.1016/S0007-8506(07)60185-1
5. T. Masuzawa, State of the art of micromachining. CIRP Ann. Manuf. Technol. **49**(2), 473–488 (2000). https://doi.org/10.1016/S0007-8506(07)63451-9
6. M.J. Madou, Fundamentals of microfabrication: the science of miniaturization, in *Fundamentals of Microfabrication: The Science of Miniaturization*, vol. 49 (2002). https://doi.org/10.1038/nmat2518
7. J. Corbett, R.A. McKeown, G.N. Peggs, R. Whatmore, Nanotechnology: international developments and emerging products. CIRP Ann. Manuf. Technol. **49**(2), 523–545 (2000). https://doi.org/10.1016/S0007-8506(07)63454-4
8. E.B. Brousseau, S.S. Dimov, D.T. Pham, Some recent advances in multi-material micro- and nano-manufacturing. Int. J. Adv. Manuf. Technol. **47**(1–4), 161–180 (2010). https://doi.org/10.1007/s00170-009-2214-5
9. S.S. Dimov, C.W. Matthews, A. Glanfield, P. Dorrington, A roadmapping study in multi-material micro manufacture, in *4M 2006—Second International Conference on Multi-Material Micro Manufacture* (2006), pp. xi–xxv. https://doi.org/10.1016/B978-008045263-0/50001-5
10. Y. Qin, A. Brockett, Y. Ma, A. Razali, J. Zhao, C. Harrison, D. Loziak, Micro-manufacturing: research, technology outcomes and development issues. Int. J. Adv. Manuf. Technol. **47**(9–12), 821–837 (2010). https://doi.org/10.1007/s00170-009-2411-2
11. M. Vaezi, H. Seitz, S. Yang, A review on 3D micro-additive manufacturing technologies. Int. J. Adv. Manuf. Technol. (2013). https://doi.org/10.1007/s00170-012-4605-2
12. X. Liu, R.E. DeVor, S.G. Kapoor, K.F. Ehmann, The mechanics of machining at the microscale: assessment of the current state of the science. J. Manuf. Sci. Eng. **126**(4), 666 (2004). https://doi.org/10.1115/1.1813469
13. S.M. Spearing, Materials issues in microelectromechanical systems (MEMS). Acta Mater. **48** (1), 179–196 (2000). https://doi.org/10.1016/S1359-6454(99)00294-3
14. K. Liu, B. Lauwers, D. Reynaerts, Process capabilities of micro-EDM and its applications. Int. J. Adv. Manuf. Technol. **47**(1–4), 11–19 (2010). https://doi.org/10.1007/s00170-009-2056-1
15. K.P. Rajurkar, G. Levy, A. Malshe, M.M. Sundaram, J. McGeough, X. Hu, A. DeSilva, Micro and nano machining by electro-physical and chemical processes. CIRP Ann. Manuf. Technol. **55**(2), 643–666 (2006). https://doi.org/10.1016/j.cirp.2006.10.002
16. X.M. Hu, Photolithography technology in electronic fabrication, in *International Power, Electronics and Materials Engineering Conference*, (Ipemec) (2015), pp. 849–856
17. N. Taniguchi, *Nanotechnology: Integrated Processing Systems for Ultra-Precision and Ultra-Fine Products* (University Press, Oxford, 1996)
18. A.P.G. Robinson, R. Lawson, *Materials and Processes for Next Generation Lithography. Frontiers of Nanoscience*, vol. 11 (2016). https://doi.org/10.1016/B978-0-08-096355-6.00001-5
19. J. Van Schoot, H. Schift, next-generation lithography—an outlook on EUV projection and nanoimprint. Adv. Opt. Technol. (2017). https://doi.org/10.1515/aot-2017-0040
20. V.Y. Banine, J.P.H. Benschop, H.G.C. Werij, Comparison of extreme ultraviolet sources for lithography applications. Microelectron. Eng. **53**(1), 681–684 (2000). https://doi.org/10.1016/S0167-9317(00)00404-4
21. R. Lebert, L. Aschke, K. Bergmann, S. Düsterer, K. Gäbel, D. Hoffmann, C. Ziener, Preliminary results from key experiments on sources for EUV lithography. Microelectron. Eng. **57–58**, 87–92 (2001). https://doi.org/10.1016/S0167-9317(01)00533-0
22. S.R. Mohanty, C. Cachoncinlle, C. Fleurier, E. Robert, J.M. Pouvesle, R. Viladrosa, R. Dussart, Recent progress in EUV source development at GREMI, in *Microelectronic Engineering*, vol. 61–62 (2002), pp. 179–185. https://doi.org/10.1016/S0167-9317(02)00572-5

23. S. Ohki, S. Ishihara, An overview of X-ray lithography. Microelectron. Eng. **30**(1–4), 171–178 (1996). http://www.sciencedirect.com/science/article/B6V0W-4287HCF-3/2/e88a050eed138b8d9ca2af46b71c5911
24. N. Mojarad, J. Gobrecht, Y. Ekinci, Interference lithography at EUV and soft X-ray wavelengths: principles, methods, and applications. Microelectron. Eng. **143**, 55–63 (2015). https://doi.org/10.1016/j.mee.2015.03.047
25. Y. Desta, J. Goettert, X-Ray masks for LIGA microfabrication, in *LIGA and its Applications*, vol. 7 (2009), pp. 11–50. https://doi.org/10.1002/9783527622573.ch2
26. Y. Cheng, B.Y. Shew, M.K. Chyu, P.H. Chen, Ultra-deep LIGA process and its applications, in *Nuclear Instruments and Methods in Physics Research, Section A: Accelerators, Spectrometers, Detectors and Associated Equipment, 467–468* (PART II), pp. 1192–1197 (2001). https://doi.org/10.1016/S0168-9002(01)00606-4
27. Y. Hirata, LIGA process—micromachining technique using synchrotron radiation lithography—and some industrial applications, in *Nuclear Instruments and Methods in Physics Research, Section B: Beam Interactions with Materials and Atoms,* vol. 208 (2003), pp. 21–26. https://doi.org/10.1016/S0168-583X(03)00632-3
28. C.K. Malek, V. Saile, Applications of LIGA technology to precision manufacturing of high-aspect-ratio micro-components and -systems: a review. Microelectron. J. (2004). https://doi.org/10.1016/j.mejo.2003.10.003
29. S. Okazaki, High resolution optical lithography or high throughput electron beam lithography: the technical struggle from the micro to the nano-fabrication evolution. Microelectron. Eng. (2015). https://doi.org/10.1016/j.mee.2014.11.015
30. L.R. Harriott, S.D. Berger, C. Biddick, M.I. Blakey, S.W. Bowler, K. Brady, D. Windt, The SCALPEL proof of concept system. Microelectron. Eng. **35**(1–4), 477–480 (1997)
31. Y. Chen, Nanofabrication by electron beam lithography and its applications: a review. Microelectron. Eng. (2015). https://doi.org/10.1016/j.mee.2015.02.042
32. R.A. Lawes, Future trends in high-resolution lithography. Appl. Surf. Sci. **154**, 519–526 (2000). https://doi.org/10.1016/S0169-4332(99)00478-X
33. J. Chae, S.S. Park, T. Freiheit, Investigation of micro-cutting operations. Int. J. Mach. Tools Manuf. **46**(3–4), 313–332 (2006). https://doi.org/10.1016/j.ijmachtools.2005.05.015
34. T. Sumitomo, H. Huang, L. Zhou, J. Shimizu, Nanogrinding of multi-layered thin film amorphous Si solar panels. Int. J. Mach. Tools Manuf. **51**(10–11), 797–805 (2011). https://doi.org/10.1016/j.ijmachtools.2011.07.001
35. D. Huo, *Micro-Cutting: Fundamentals and Applications* (Wiley, London, 2013)
36. G. Byrne, D. Dornfeld, B. Denkena, Advancing cutting technology. CIRP Ann. Manuf. Technol. **52**(2), 483–507 (2003). https://doi.org/10.1016/S0007-8506(07)60200-5
37. N. Ikawa, R.R. Donaldson, R. Komanduri, W. König, T.H. Aachen, P.A. McKeown, I.F. Stowers, Ultraprecision metal cutting—the past, the present and the future. CIRP Ann. Manuf. Technol. **40**(2), 587–594 (1991). https://doi.org/10.1016/S0007-8506(07)61134-2
38. S. Shabouk, T. Nakamoto, Micro machining of diamond by ferrous material, in *MHS 2001—Proceedings of 2001 International Symposium on Micromechatronics and Human Science* (2001), pp. 35–40. https://doi.org/10.1109/MHS.2001.965218
39. J. Xie, M.J. Luo, K.K. Wu, L.F. Yang, D.H. Li, Experimental study on cutting temperature and cutting force in dry turning of titanium alloy using a non-coated micro grooved tool. Int. J. Mach. Tools Manuf. **73**, 25–36 (2013). https://doi.org/10.1016/j.ijmachtools.2013.05.006
40. H.N. Li, T.B. Yu, Z.X. Wang, L.Da Zhu, W.S. Wang, Detailed modeling of cutting forces in grinding process considering variable stages of grain-workpiece micro interactions. Int. J. Mech. Sci. **126**, 319–339 (2017). https://doi.org/10.1016/j.ijmecsci.2016.11.016
41. H.H. Gatzen, J. Chris Maetzig, Nanogrinding. Prec. Eng. **21**, 134–139 (1997). https://doi.org/10.1016/S0141-6359(97)00082-2
42. M. Hasan, J. Zhao, Z. Jiang, A review of modern advancements in micro drilling techniques. J. Manuf. Process. (2017). https://doi.org/10.1016/j.jmapro.2017.08.006

43. S. Kumar, A. Dvivedi, P. Kumar, On tool wear in rotary tool micro-ultrasonic machining, in *Minerals, Metals and Materials Series* (2017), pp. 75–82. https://doi.org/10.1007/978-3-319-52132-9_8
44. K.P. Rajurkar, M.M. Sundaram, Process improvements in micro USM and micro EDM. *Micro*, Dec 2015
45. Z. Yu, C. Ma, C. An, J. Li, D. Guo, Prediction of tool wear in micro USM. CIRP Ann. Manuf. Technol. **61**(1), 227–230 (2012). https://doi.org/10.1016/j.cirp.2012.03.060
46. D.V. Srikanth, M.S. Rao, Abrasive jet machining—research review. Int. J. Adv. Eng. Technol. **5**(2), 18–24 (2014)
47. O.Y. Rogov, V.V. Artemov, M.V. Gorkunov, A.A. Ezhov, S.P. Palto, Fabrication of complex shape 3D photonic nanostructures by FIB lithography, in *IEEE-NANO 2015—15th International Conference on Nanotechnology* (2015), pp. 136–139. https://doi.org/10.1109/NANO.2015.7388897
48. G. D'Urso, C. Merla, Workpiece and electrode influence on micro-EDM drilling performance. Prec. Eng. **38**(4), 903–914 (2014). https://doi.org/10.1016/j.precisioneng.2014.05.007
49. M.P. Jahan, M. Rahman, Y.S. Wong, Micro-electrical discharge machining (Micro-EDM): processes, varieties, and applications, in *Comprehensive Materials Processing*, vol. 11 (2014), pp. 333–371. https://doi.org/10.1016/B978-0-08-096532-1.01107-9
50. J. Fleischer, T. Masuzawa, J. Schmidt, M. Knoll, New applications for micro-EDM. J. Mat. Process. Technol. **149**, 246–249 (2004). https://doi.org/10.1016/j.jmatprotec.2004.02.012
51. D. Pham, S. Dimov, S. Bigot, A. Ivanov, K. Popov, Micro-EDM—recent developments and research issues. J. Mater. Process. Technol. **149**(1–3), 50–57 (2004). https://doi.org/10.1016/j.jmatprotec.2004.02.008
52. Z. Wang, Y. Zhang, W. Zhao, Study on key technologies of micro-EDM equipment, in *Proceedings 4th Euspen International Conference*. Glasgow, Scotland, May–June 2004, pp. 51–52
53. A. Moarrefzadeh, Study of electron generation in electron beam machining (EBM) process. Int. Rev. Mech. Eng. **5**(6), 1064–1070 (2011)
54. S. Gao, H. Huang, Recent advances in micro- and nano-machining technologies. Front. Mech. Eng. (2017). https://doi.org/10.1007/s11465-017-0410-9
55. K.H. Choi, J. Meijer, T. Masuzawa, D.H. Kim, Excimer laser micromachining for 3D microstructure. J. Mat. Process. Technol. **149**, 561–566 (2004). https://doi.org/10.1016/j.jmatprotec.2004.03.005
56. N.H. Rizvi, P. Apte, Developments in laser micro-machining techniques. J. Mater. Process. Technol. **127**(2), 206–210 (2002). https://doi.org/10.1016/S0924-0136(02)00143-7
57. J. Meijer, K. Du, A. Gillner, D. Hoffmann, V.S. Kovalenko, T. Masuzawa, W. Schulz, Laser machining by short and ultrashort pulses, state of the art and new opportunities in the age of the photons. CIRP Ann. Manuf. Technol. **51**(2), 531–550 (2002). https://doi.org/10.1016/S0007-8506(07)61699-0
58. D. Mi, W. Natsu, Simulation of micro ECM for complex-shaped holes, in *Procedia CIRP*, vol. 42 (2016), pp. 345–349. https://doi.org/10.1016/j.procir.2016.02.186
59. Drexler, K. E. (1992). *Nanosystems: Molecular machinery, manufacturing and computation. Advanced Materials* (Vol. 5). https://doi.org/10.1002/adma.19930051119
60. T.L. Cocker, V. Jelic, M. Gupta, S.J. Molesky, J.A.J. Burgess, G.D.L. Reyes, F.A. Hegmann, An ultrafast terahertz scanning tunnelling microscope. Nat. Photon. **7**(8), 620–625 (2013). https://doi.org/10.1038/nphoton.2013.151
61. T.V. Vorburger, J.A. Dagata, G. Wilkening, K. Iizuka, Industrial uses of STM and AFM. Comput. Stand. Interf. **21**(2), 196 (1999). https://doi.org/10.1016/S0920-5489(99)92292-4

Part II
Micro and Nano Machining with Conventional Machining Techniques

Micromachining of Titanium Alloys

Md. Rashef Mahbub and Muhammad P. Jahan

1 Introduction

High strength-to-weight ratio, excellent corrosion resistance, and biocompatibility have made Titanium (Ti) and its alloys one of the most popular group of materials in modern industrial processes. It is widely used in aerospace, chemical plants, automobile, power generation, oil and gas extractions, and biomedical applications [1]. Two major applications of titanium alloys are in aerospace and biomedical industries.

Commercially, pure titanium has been classified into four main grades according to their difference in mechanical properties. Grade 1 is the softest and most ductile in the group, which is generally popular in chemical processing, chlorate manufacturing, dimensionally stable anodes, desalination, architecture, medical industry, marine industry, automotive parts, and airframe structure. The main characteristic of this grade is formability. It is also known for its excellent corrosion resistance and high impact toughness. The second grade of titanium is one of the most popular among pure titanium grades and slightly stronger than grade 1. It has excellent weldability, strength, ductility, and formability. For the superior mechanical properties, grade 2 titanium alloy is widely used in architecture, power generation, medical industry, hydrocarbon processing, marine industry, exhaust pipe shrouds, airframe skin, desalination, chemical processing, and chlorate manufacturing. Though grade 3 is stronger than both grade 1 and grade 2, it is the least used among all titanium grades and its major uses are in aerospace structures, chemical processing, medical, and marine industries. Grade 4 is the strongest of the titanium grades and also known as medical grade titanium. It is known for its formability, weldability, and excellent corrosion resistance. Industrial applications where high

Md. Rashef Mahbub · M. P. Jahan (✉)
Department of Mechanical and Manufacturing Engineering, Miami University, Oxford, OH 45056, USA
e-mail: jahanmp@miamiOH.edu

K. Kumar et al. (eds.), *Micro and Nano Machining of Engineering Materials*, Materials Forming, Machining and Tribology,
https://doi.org/10.1007/978-3-319-99900-5_3

strength is needed, grade 4 titanium alloy is one of the prime choices. Excellent mechanical strength made it very popular for the applications in airframe components, cryogenic vessels, heat exchangers, CPI equipment, condenser tubing, surgical hardware, and pickling baskets [2]. Ti-6Al-4V or grade 5 titanium alloy is probably the most popular among all titanium alloys because of its lightweight compared to high strength, corrosion resistance, and biocompatibility. It is widely used in many manufacturing applications but is most popular in biomedical and aerospace industries. Ti5Al2.5Sn, grade 6, is more commonly used for airframe and jet engine applications requiring good weldability, stability, and strength at elevated temperatures. The inclusion of palladium makes grade 7 titanium an alloy which has pretty similar characteristics with grade 2. It has good weldability and fabricability, but it is more popular for its corrosion resistance as it is most corrosion resistant among all titanium alloys. It is widely used in chemical processing and production equipment component. Grade 11 titanium has similar characteristics as grade 1, but a small inclusion of palladium for corrosion resistance makes it an alloy. It can reduce acid in a chloride environment and is used mainly in applications where corrosion is a main issue. Apart from these, there are also different alloys like grade 23, grade 12, etc., which are being used for different purposes [2].

Although titanium and its alloys are superior in many characteristics, some other properties have made it really a difficult-to-machine material. Tool life, material removal rate, cutting forces, surface finish or chip morphology are some of the main parameters to be considered while judging the machinability of a material, and titanium can be defined as a material of poor machinability in this regard. Being a poor thermal conductor, titanium cannot dissipate most of the heat generated in machining process properly, rather most part of this heat gets concentrated on cutting edge and tool face. While machining, the machining environment generates very high temperature, and titanium being chemically very reactive exhibits strong alloying tendency and starts to do so with tool material at this high temperature resulting in galling, welding, and smearing. Because of the thermal plastic instability in titanium, the shear strains gets localized in a narrow band which results in serrated chip formation and there remains an unusual short contact length between the chip and the cutting tool. Serrated chips pave the way in fluctuating the cutting force which is also very dangerous while machining. Extreme flank wear is noticed because of micro-fatigue loading of cutting tool, which results in vibrational force coupled with high temperature [3].

Based on the literature, conventional micromachining processes, such as micro-turning, micro-milling, micro-grinding, micro-drilling, etc., have been found to be challenging for machining most of the titanium alloys. On the other hand, nonconventional micromachining processes, such as micro-electro-discharge machining (micro-EDM), micro-electrochemical machining (micro-ECM), laser micro-micromachining, abrasive water-jet micromachining, etc., are being approached more extensively in recent years for micromachining of titanium alloys. Some challenges in the area of conventional micromachining of titanium alloys that need future considerations are reactivity of titanium alloys to cutting tool materials, burr formation, and adhesion to the tool, tool wear, surface finish, size effect, and limitation to miniaturization due to cutting forces. The challenges in nonconventional micromachining of titanium alloys are comparatively lower to material removal rate, poor surface integrity, recast layer formation, heat-affected zone, and changes in microstructures and mechanical properties of the part after machining. All the challenges in nonconventional and conventional micromachining with possible solutions to overcome the challenges will be discussed in the chapter.

This chapter presents an overview of various micromachining processes to machine titanium and titanium alloys by discussing the processes and related research works on those fields. It is divided into two main sections: conventional and nonconventional micromachining of titanium alloys. Upon discussing all these micromachining processes, the challenges while machining titanium alloys at microscale has been identified, and future works that might be done to resolve those issues have been proposed.

2 Conventional Micromachining of Titanium Alloys

2.1 Micro-Milling of Titanium Alloys

Micro-end milling process is one of the most versatile micromachining process available because of its flexibility. It is vastly used in medical and aerospace industries because of its ability to produce complex parts. However, while machining in microscale there are a lot of criterions that have to be taken into consideration which can easily be ignored in conventional and macroscale. Things like vibration, deflection, temperature, microstructure of workpiece, etc., become very important while machining using micro-milling, especially when machining hard-to-machine materials like titanium and its alloys at microscale. As surface topography of micro-features is in submicrometer order, it is difficult to achieve a smooth surface in micro-milling by just finishing process. But this is one of the most important performance criteria of micromachining process as surface topography of machined surface controls or manipulates other criteria like vibration, lubrication, biocompatibility, etc., to some great extent. To get better surface finish in micro-milling process, measures like optimization of parameters, modeling, and simulation of surface generation, use of different lubrication have been tried time to time.

Burr formation is one of the major issues that has to be dealt with while micromachining of titanium alloys. Burr formation process is primarily dependent on material's property like ductility, cutter geometry, cutting parameters, tool wear, and shape of workpiece. It is even more prominent while machining hard materials like titanium because of tool wear. Burr formation is almost impossible to avoid in micromachining processes like micro-milling, but optimum process parameter selection might minimize this phenomenon to a bearable limit. However, achieving a parameter setting that does justice for both, minimizing surface roughness and burr formation is really a challenging one as one parameter setting that works for one of

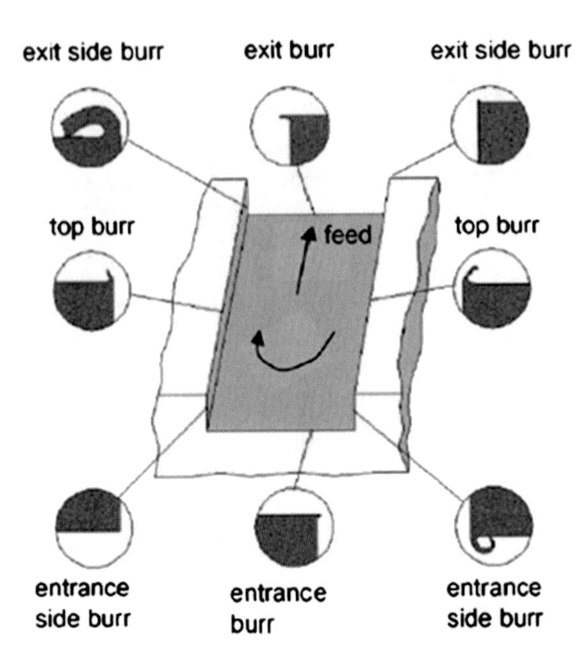

Fig. 1 Different kind of burrs formed during the micro-milling of titanium alloys [4]

these two might not work for the other one [4]. Figure 1 presents different types of milling burrs typically formed during the micro-end milling of titanium alloys [4]. The graphs shown in Fig. 2 indicates the relation of surface roughness and burr formation with the selected process parameters. It was found from the experiments that axial depth of cut plays the most important role in surface roughness and it was the feed per tooth/feed rate which affects the surface roughness most. Though the relation of feed rate with surface roughness here is opposite as higher feed rates deliver better surface finish and better quality of micro-channels [4].

Finite element analysis (FEA) simulation has become an important tool in different machining processes. The main advantage of FEA simulation is that it can computationally predict different kind of performance criterions or effects of machining parameters on the cutting temperature, strain, strain rate, stress, etc., which are otherwise hard to determine using experimental method [5]. Specially while micromachining, it becomes extremely difficult to reach out every part to get detail information and that is why FEA is even more important and useful in the analysis of micromachining processes. Orthogonal or 2D finite element (FE) model is easier than the 3D model to simulate and a good number of research work has been conducted on this field. Damage criteria based serrated chip formation simulations have been conducted using Ti-6Al-4V as machining material [6]. Another 2D model of the same alloy has been designed utilizing flow softening behavior of the material [7]. Study has been done to modify adiabatic shearing based serrated chip formation simulation [8]. All these studies helped a lot for better understanding of the physics of micro-milling of titanium alloys. The oblique or 3D model obviously can predict the effects of machining parameters on the machining

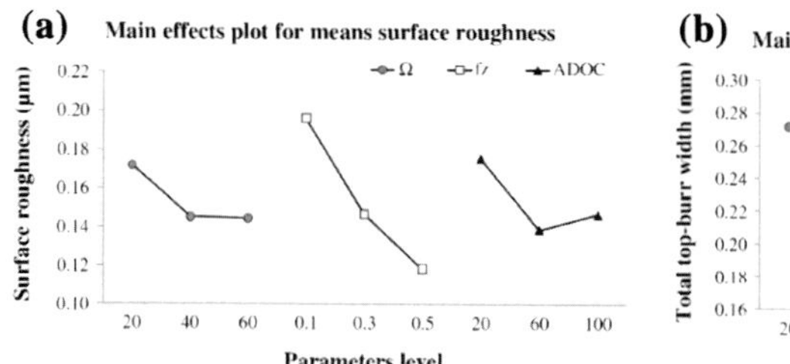

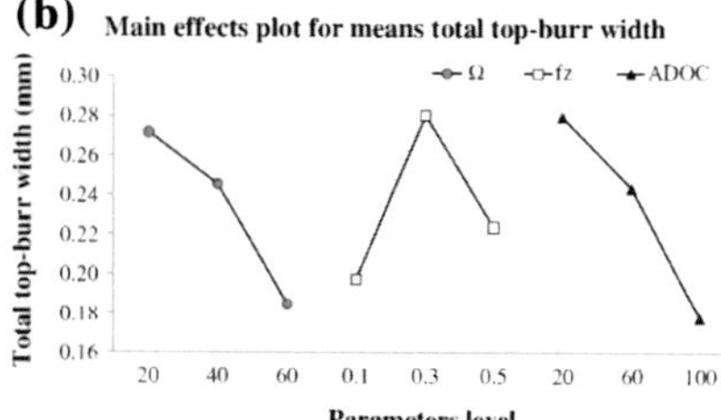

Fig. 2 Main effects plots for surface roughness and top-burr width during micro-milling of Ti-6Al-4V, where Ω = spindle speed in thousands revolution per minute, f_z = feed per tooth in micrometer, ADOC = axial depth of cutin micrometer [4]

performance more accurately and gives a better understanding of the whole process. A 3D FE model and simulation has been conducted for micro-end milling of Ti-6Al-4V and was validated with experimental data in terms of chip flow and tool wear. Effect of edge roundness from tool wear development has also been investigated [5]. Figure 3 shows the FE simulation of solid cutting tool. Figure 4 shows the geometry of a micro-end mill cutter showing speed (π), feed per tooth (fz), and axial depth of cut (a_p) as three main process parameters in micro-milling.

Three different types of micro-end milling process during micromachining of Ti-6Al-4V were modeled [5], which are full-immersion slot micro-end milling, half-immersion up micro-end milling and half-immersion down micro-end milling. For full-immersion 2×10^5 elements and for half-immersion 1.5×10^5 elements were considered for meshing where the minimum element size was 0.5 μm. Figure 5 shows the FEA simulation of three types of micro-end milling process [5]. During the FEA analysis, sticking and sliding contact were considered along with the tool-workpiece contact length. It was reported that the edge radius was

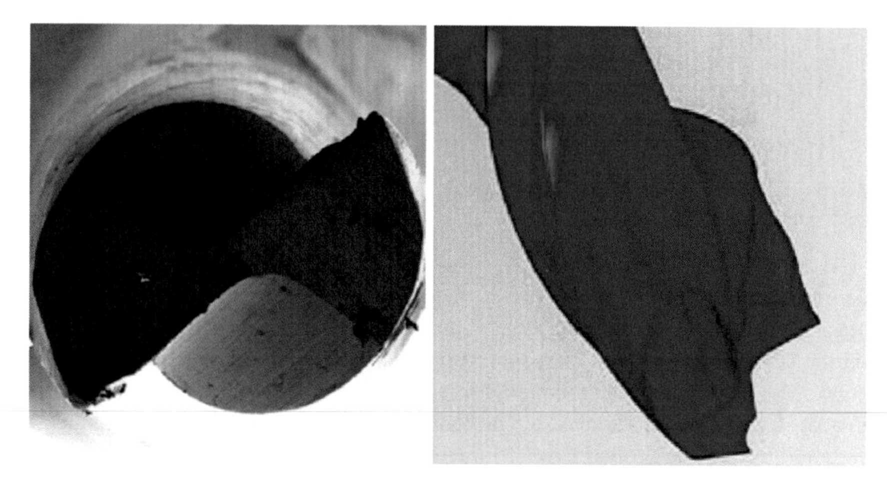

Fig. 3 SEM image of micro-end milling tool [5]

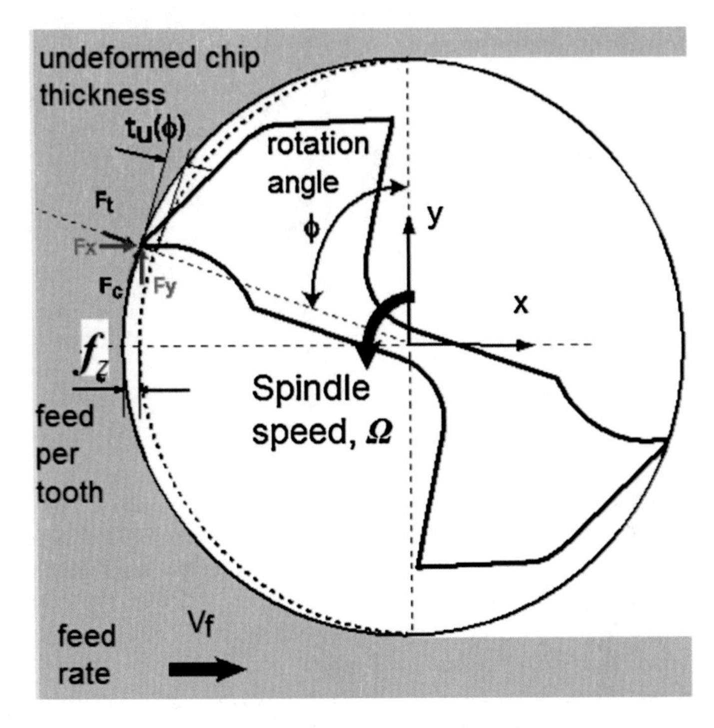

Fig. 4 Micro-end milling process parameters on full-immersion [5]

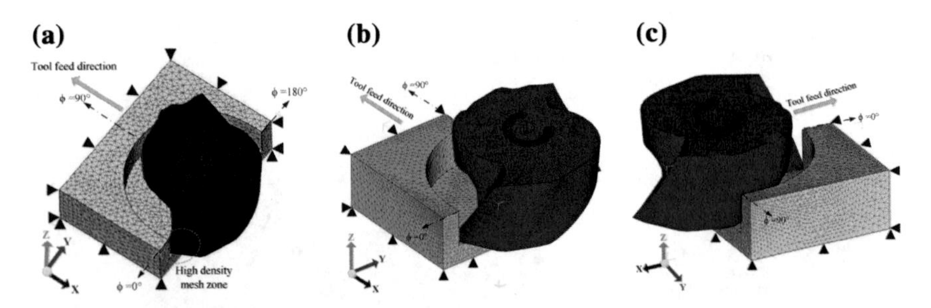

Fig. 5 3D FE model for **a** full-immersion slot, **b** half-immersion up, **c** half-immersion down micro-end milling [5]

comparably large in micromachining than uncut chip thickness which resulted in negative rake angle. Increasing edge radius due to tool wear makes it even higher on the negative side which ends up creating a smaller shear angle and larger shear zone. More force is then required for plastic deformation and temperature also increases at the cutting zone [5]. The tool wear increases with increasing value of edge radius and edge roundness also increase because of tool wear. Wearing out of tool is not a constant phenomenon rather an exponential one. While tool wear reaches a certain level, built up edge might be formed. It compensates the influence of larger negative rake angle. It was found that ploughing was less in down milling than up milling. So in case of micro-milling, down milling provides more sustainability than up milling and more suitable for applications where long cutting duration is required [5].

2.2 *Micro-Drilling of Titanium Alloys*

In recent years, micromachining has found applications in various industries including electronic, medical, automobile, and aerospace industries. Different kind of miniaturized products like PCB (printed circuit board), microscopic nozzles, micro dies and molds, chemical microreactors, tooth implantation, high-tech medical appliances, fuel filters, and fuel ignition systems are being produced in a large scale with the help of micromachining techniques. Micro-drilling is one of the most popular and important processes among those micromachining techniques which basically implies the drilling of diameters between 1 μm and 1 mm [9]. Micro-drilling process is extensively used in precision engineering, micro-electromechanical systems (MEMS), micro total analysis (μTAS), consumer products, biomedical and chemical engineering, optical displays, fluidics, wireless and optical communications, and PCB industries [10–12]. The most common categories of conventional mechanical micro-drilling are twist type, spade, D-shaped, single flute, compound tool micro-drilling, coated micro-drilling, etc. The most popular nonconventional techniques are laser, electric discharge machining, electrochemical machining, spark-assisted chemical engraving micro-drilling, electron beam micro-drilling, micro-drilling by ultrasonic vibration, etc. [9].

Abrasive machining technique is one of the popular methods applied to fabricate micro-drills. To fabricate micro-drills of microscale size, the types of machine tools require specialty as well. Precise and accurate measurement is needed throughout the whole system and grinding wheels are also required to be ultra-fine. At first, a rod is cut into sections according to the dimension needed. To give the outer diameter of the blank rod a shape, centerless grinding is performed. While cutting the rod in sections, burr formation becomes one of the main problems and that is why profile grinding is done. Chamfering is also done with chamfer angle generally 45°–60°. With point angle grinding, drill point is created then. A special type of ultra-fine grinding wheel is then used to grind cut the cutting part known as body length. Flutes are made in next step and edges are made, sharpened in flank area for producing cutting edges of micro-drill. A cleaning process is then conducted that flush away oil and dust and at last dimensions are measured. The main disadvantage of this process is the huge time that is needed in machining process. It requires quite a few numbers of different machine and tools so this process is not cost friendly as well for high maintenance labor cost. Using forming method instead of abrasive one might be a solution to this problem. A forming technique that will be able to solidify the powder simultaneously giving the product its final shape can be used.

Another option is that with the negative replica of micro-drill, microdie can be made and a forming process like extrusion needs to be performed [9].

Now, with the advancement of technology, microdrill less than a diameter of human hair can be produced as well. Though in commercial section, size, shape, and materials are main concern, research section is concerned with a whole lot of other performance improving parameters like mechanical properties [13, 14], simulation and numerical analysis [15, 16], micro-drill parameter optimization [10, 17], surface texture of drills [18], chip formation analysis [19, 20], lubrication methods [21, 22], inspection mechanisms [23, 24], re-sharpening processes [25, 26], perforation techniques in different types of specimen [12, 27], effects of coating [28, 29], and prevention of tool breakage [30, 31].

The high cutting force during micro-drilling of titanium alloys causes vibration in the spindle axis which results in poor surface quality. Tool life is also reduced while dealing with higher cutting force generated during machining of titanium alloys. Large torque also creates high temperature on tool–workpiece interface because of increasing friction. It is reported that titanium drilling generates higher cutting force than drilling aluminum and steel [32, 33]. Also, much higher stress occurs at cutting edge while micro-drilling of titanium alloy (Ti-6Al-4V) than machining other nickel-based alloys or steel. Titanium alloys show strong resistance at elevated temperature and the chip-tool contact area is also ridiculously small on rake face. That might be the reason behind higher stress. Because of poor thermal conductivity of titanium alloys, majority (around 80%) of the heat produced from plastic deformation while micro-drilling is absorbed by the tool. This is pretty lower (50–60%) in case of steel. This is the reason why rapid tool wear is a common incident while drilling titanium alloys at microscale.

2.3 *Micro-Turning of Titanium Alloys*

Micromachining can be classified into two basic groups called mask-based and tool-based [34]. Micro-turning is one of the common tool-based micromachining processes. In this process, the workpiece and the cutting tool need to be moved relative to each other for chip removal. Following various cutting path, 3D shape on microscale can be produced by this process [35]. Cutting speed, feed rate, and depth of cut are the three main parameters considered in the micro-turning process. During micro-turning process, a certain cutting speed is maintained with the primary motion and lateral speed of the tool is much slower comparably which is called the feed. The height or length cutting tool penetrates below the work surface is known as depth of cut in this process. This dominant cutting force continues to remove material in the form of chips and newly removed chips continue to expose new to be machined surface [36].

Most of the machinable material can be machined using micro-turning. However, during the machining, workpiece gets deflected time to time by thrust force which is one of the main drawbacks of micro-turning process [37]. Study has been progressed on different machining parameters, such as spindle speed, feed rate, depth of cut, etc., and their influences on performance criterions like tool wear and surface roughness during micro-turning of titanium alloys [37, 38]. The major parameters that affect the performance of the micro-turning process are presented with the fish-bone diagram, as shown in Fig. 6. Generally cutting forces increase with the increasing of cutting speed and feed. From the metallurgical study of grain size and density of titanium alloy after machining, it is advised that selecting material from available standard data is not a very good idea as this process in micro level are extremely precision depending and little deviation or inaccuracy might lead to major difference in machinability and precision [38].

In recent years, dry micro-turning of titanium alloys has been quite popular where no coolant is used. Uncoated and CVD-coated carbide tools are being used in micro-turning of titanium alloy [39]. But tool chip adhesion does create some issues in this method. As a result, for decreasing friction on tool rake surface, microscale texture on the cutting tool was suggested. Micro-grooved tool of 0.1–0.15 μm too nose diameter were able to reduce chip adhesion to tool surface while machining titanium alloys in wet cutting condition, but it has opposite effect in dry cutting [40]. But while using tool of 0.5–1.2 μm, tool nose diameter tools with TiN coating, it was able to decrease chip adhesion in both cutting conditions [41]. Micro-grooved tools of different tool nose diameters have been used on rake surface while dry machining of titanium alloy to investigate the influence of micro-grooves shape and size on cutting force and tool tip temperature. It was found that compared to the traditional plane tool, micro-grooved tool can decrease cutting temperature to a great extent, almost a 100 °C during micro-turning of titanium alloy. Cutting tool tip temperature was also found to be more than double in traditional plane tool than the micro-grooved one. It is also found that decreasing groove depth increases shear angle and decreases cutting tool temperature [42]. From SEM micro-topographies of cutting chips it is found that micro-grooved tool promotes more stable cutting condition than the traditional plane tool by creating finer sawtooth on-chip free surface (Fig. 7).

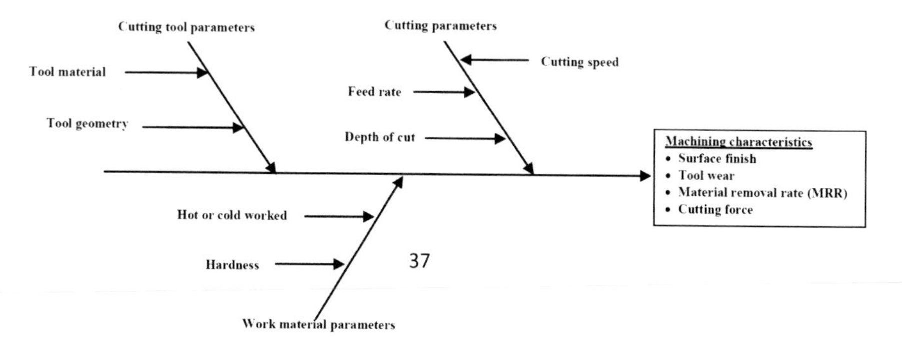

Fig. 6 Process parameters related to performance criterions during micro-turning [36]

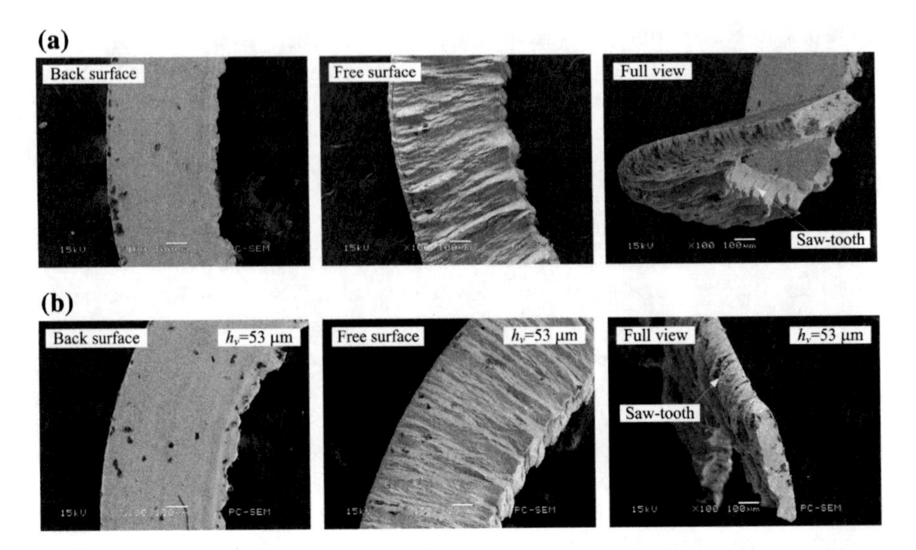

Fig. 7 Micro-topographies of Ti-6Al-4V chips during micro-turning using **a** traditional plane tool, **b** micro-grooved tool [42]

2.4 Micro-Grinding of Titanium Alloys

A grinding wheel of microscale, which is generally made of diamond or aluminum abrasive particles are used in micro-grinding process for finishing operations. Those particles are embedded in a solid metal matrix on the grinding wheel. The smoothness of the machined surface depends greatly upon the fineness of those particles. The finer the particle is the smoother the machined surface finish will be in micro-grinding operation.

Very few research has been conducted on micro-grinding of titanium alloys. Investigation on micro-grinding of Ti-6Al-4V has been carried out with two different grinding wheels: vitrified silicon carbide (SiC) wheel and resin bonded cubic boron nitride (CBN) wheels. A grindable thermocouple was used to record the temperature response on the workpiece surface. Scanning electron microscopy (SEM) and energy dispersive spectroscopy (EDS) were used to investigate the machined surface morphologies of the workpiece and the worn out wheel surface. At high temperature, strong welds tend to form between Ti-6Al-4V and SiC. Existing chemical bonding is the main reason behind this. This bonding force exceeding the shear strength of the metal creates an adhering layer on the abrasive grit. Gradually, a layered structured seems to be formed on the worn abrasive grit. This adhering layer growing large enough starts to act as a cantilever and eventually goes in the verge of losing it. Metal particles redepositing on the workpiece forms a plastically deformed coating and sometimes looks like a scar. Mechanical characteristics of materials are also responsible behind this apart from the chemical factors. Titanium alloys become more ductile at high temperature and, therefore,

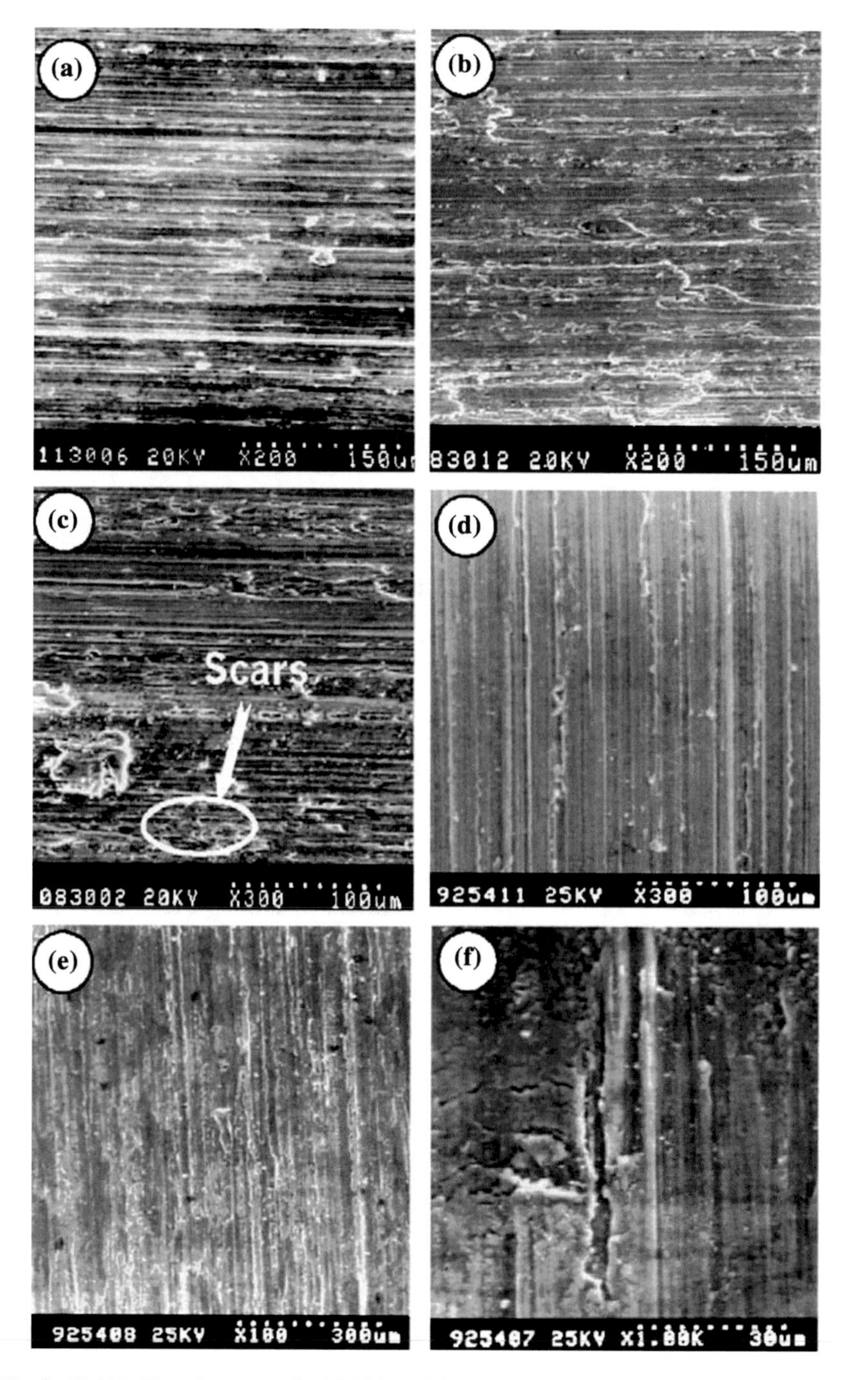

Fig. 8 Ti-6Al-4V surface ground with SiC and CBN wheels (SEM micrograph). SiC grinding when temperature is **a** 100 °C, **b** 500 °C and **c** 1000 °C. CBN grinding when temperature is **a** 80 °C, **b** 600 °C and **c** 1300 °C [43]

cold-welding occurs between the workpiece and abrasive grits [43]. Figure 8 shows the machined surfaces obtained after micro-grinding of Ti-6Al-4V using SiC and CBN wheels at different temperatures [43].

3 Nonconventional Micromachining of Titanium Alloys

3.1 Micro-Electrical Discharge Machining of Titanium Alloys

In micro-electrical discharge machining (micro-EDM), high voltage is applied at the gap between electrode and the workpiece submerged in a dielectric medium creating a series of sparks between the workpiece and electrode. This series of sparks creates localized heated zone and both workpiece and electrode materials start melting and vaporizing. Swiftly solidified ablated material is driven away from the electrode gap in the form of debris by the dielectric fluid used in the process. Thus, the material is removed from the workpiece and electrode in the micro-EDM process [9]. The amount of material removal from the workpiece and electrode can be controlled by selecting and interchanging the polarity. As a noncontact process, micro-EDM is suitable for machining most of the electrically conductive materials, specially for creating a part with complex geometries, while dealing with hard-to-machine materials like titanium, nickel, and their alloys.

Parametric studies of micro-EDM with titanium alloys Ti-6Al-4V have been conducted by many researchers. Study has been progressed to investigate material removal rate (MRR), tool wear (TW), overcut and taper taking peak current, pulse-on-time, dielectric flushing pressure, and duty ratio as parameters to be varied. It is found that pulse-on-time is the most influencing factor on MRR, TW, and overcut, whereas peak current contributes most while defining taper. MRR and TW both generally increase with increasing peak current and pulse-on-time, though flushing pressure and duty ratio do not have any noticeable effect on those. Overcut also increases with increasing peak current and pulse-on time [1].

Majority of the studies on micro-EDM has been conducted using micro-EDM drilling process. While drilling by micro-EDM, gaseous bubble escapes through the narrow discharge gap from the machining environment and if the micro-hole is of shallower depth, the bubble pressure can easily drive the dielectric to flow in the discharge gap. Increasing micro-hole depth increases the viscous resistance of dielectric and with the increasing of the aspect ratio of micro-hole there comes a certain value when the balance between the pressure of bubble and viscous resistance is reached [44]. As viscous resistance is decreased in uneven gap and gaseous bubbles can escape more easily from it then the even one, the planetary movement creates an uneven gap between the wall of the hole and side of electrode and this eventually increases the achievable aspect ratio of micro-holes [45]. Though micro-EDM process is considered to be a noncontact process, the process force is not zero in the real case. During the micro-EDM process, due to the electric energy at the gap, there is an electrostatic or electromagnetic force acting at the spark gap. The pressure of bubbles generated by evaporation and dissociation of the dielectric liquid creates the electrostatic force which acts on the thin long electrode, make it bent. The movement of the electrode tip is really complex which results in different shape while entrance and exit like the Fig. 9 [44]. Figure 9 shows typical profile inaccuracies caused by the electrostatic force during micro-EDM drilling of titanium alloy [44].

The dielectric used in micro-EDM mainly serves two purposes. Working as a semiconductor between the electrode and workpiece, it maintains a stable spark gap ionization condition. Second, it helps to flush away the debris from the spark gap area. The dielectric fluid used during the micro-EDM process influences the performance criterions to some great extent during micro-EDM of titanium alloys [46–48]. Material removal rate, tool wear rate, overcut, diameter variance at entry and exit hole, surface roughness, and integrity all of these are affected by the selection of dielectric fluid. Kerosene and deionized water are probably two of the most common dielectrics used during micro-EDM of titanium alloys. However, kerosene as a dielectric creates several issues as it has the tendency to deposit carbine layer on workpiece surface which eventually reduces material removal rate. It also makes the discharge inefficient as adhesion of carbon particles on micro-tool surface while using kerosene happens quite often [46]. Harmful vapors like CO and CH_4 form sometimes making the machining environment a toxic one [47, 48]. Most of these disadvantages are solved while using deionized water as dielectric fluid during micro-EDM of titanium alloys. It promotes better machining environment than kerosene as no gas vapor is formed and also no carbon deposition happens which promotes better spark discharge. It also has rapid cooling and enhanced flushing properties which eventually results in improved material removal rate and tool wear rate [49, 50] (Fig. 10).

In recent years, to improve the function of the dielectric in improving the surface finish, electrically conductive and/or semiconductive powder particles are mixed

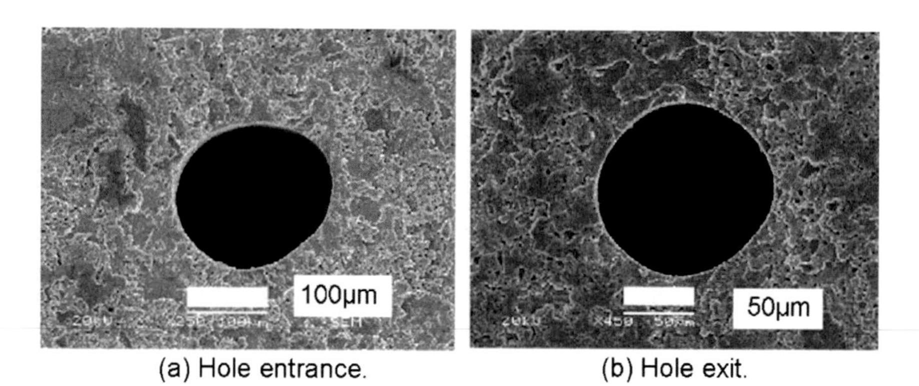

Fig. 9 Entrance and exit side of a micro-hole machined in titanium alloy by micro-EDM [44]

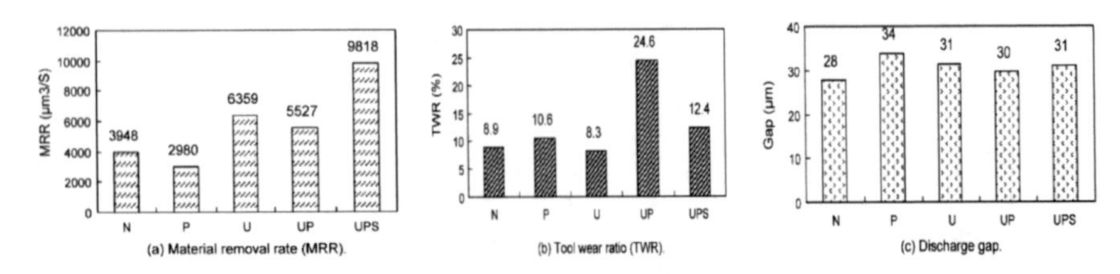

Fig. 10 Performance criterions with respect to process parameters N = normal machining Conditions, P = aided with planetary movement, U = aided with ultrasonic vibration, UPS = aided with planetary movement and ultrasonic vibration [44]

with the dielectric during the micro-EDM process. It was found that powder-mixed micro-EDM can deliver near mirror-like surface by improving surface finish and quality at relatively high machining rate during micro-EDM of titanium alloys [51–53]. Powder-mixed EDM promotes a surface that has better resistance to corrosion and abrasion [54]. The added powder helps to improve the breakdown characteristics of dielectric fluid which results in the decrease of insulating strength of the dielectric and increase of spark gap distance between the workpiece and the electrode [53, 55, 56]. This makes flushing of debris uniform promoting a stable machining environment for better surface finish and integrity. Powders of Al, Cr, Cu, graphite, silicon, and silicon carbide have been tried with kerosene and deionized water during micro-EDM of titanium alloys. There have been studies conducted dividing the type of additive in two different groups: electrically conductive and nonconductive [56–59]. Electrically conductive additives like Al, Cu, and graphite cannot withstand at high temperature being soft in nature. On the other hand, electrically nonconductive additives like SiC and Al_2O_3 are generally hard and can withstand high temperature generated during micro-EDM of titanium alloys [46]. Experiments have been conducted taking Ti-6Al-4V as workpiece material while machining micro-holes by micro-EDM process to study different performance criterions like material removal rate, tool wear, overcut, taperness, and machining time. Simple deionized water based dielectric and boron carbide (B_4C) powder-mixed deionized water were used in the process and the performance of various powders on the micro-EDM of Ti-6Al-4V was compared. Figure 11 shows the comparison of micro-holes machined by pure deionized water and B_4C powder-mixed deionized water during micro-EDM drilling of Ti-6Al-4V [46]. It is obvious from the images that B_4C mixed one promotes better surface quality and circularity both in entry and exit. Overcut was found to be more in the powder-mixed one than the pure dielectric one.

Besides improving the surface finish, the powder-mixed micro-EDM reduces the thickness of the recast layer, which is the heat-affected zone due to the electro-thermal nature of material removal in micro-EDM. Figure 12 shows the comparison of recast layer formation on Ti-6Al-4V after machined by micro-EDM using pure deionized water and B_4C mixed deionized water. It can be seen from the Fig. 12 that the thickness of recast layer was less in B_4C mixed deionized water compared to the pure deionized water. During powder-mixed micro-EDM,

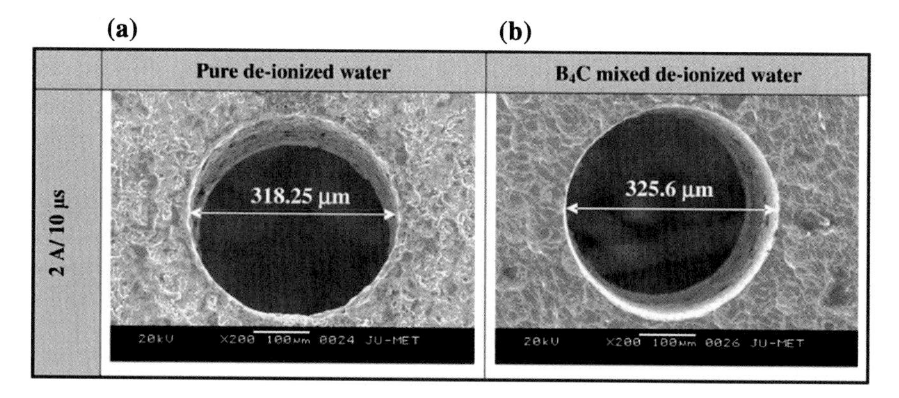

Fig. 11 Comparison of micro-holes machined on Ti-6Al-4V using micro-EDM with **a** pure deionized water and **b** B_4C power-mixed deionized water [46]

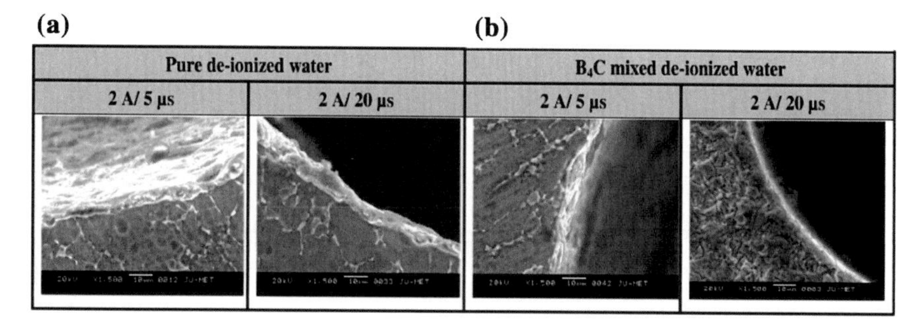

Fig. 12 Comparison of recast or white layer on micro-hole edge machined on Ti-6Al-4V by micro-EDM with **a** pure deionized water and **b** B_4C power-mixed deionized water [46]

the additive powder particles remove melted work material from the micro-hole wall limiting recast layer formation. It was found that in both cases that higher pulse-on-time creates thicker recast layer than lower pulse-on-time. In terms of material removal rate, B_4C-mixed dielectric improves the quality of discharge phenomena during micro-EDM process, thus increases the MRR compared to that provided by pure dielectric. Another added advantage of using B_4C powder-mixed dielectric during the micro-EDM of titanium alloy is the reduction of tool wear. During micro-EDM, the boron carbide powder decomposes at higher discharge energy producing adhesive carbon particles, which produces a protective coating layer on the tool electrode and reduce the tool wear.

3.2 Electrochemical Micromachining of Titanium Alloys

Electrochemical micromachining (EMM or micro-ECM) is another nonconventional process for machining hard and difficult-to-cut materials with high-quality surface finish. There is very minimal mechanical stress or tool wear and, therefore, production of heat-affected zone is not a major issue in micro-ECM. There have been research with different electrolytes, anode–cathode systems, and chemicals in this field to find out the optimum condition in different machining systems and different materials. With the advancement of technology, the electrochemical machining is being applied in microscale as well with noticeable success.

Besides high-quality surface finish, another major application of micro-ECM is in the machining of high aspect ratio micro-holes in hard and difficult-to-cut materials like titanium alloys [60, 61]. It is very difficult-to-machine micro-holes on hard and difficult-to-machine materials like titanium alloys with traditional mechanical micromachining processes, especially when dealing with critical dimensions and complex features [62]. Micro-drilling with electrochemical micromachining can be a way to solve this challenge [63]. In the micro-ECM process, an electrolyte flows between the anode and cathode which completes the whole system. Voltage is applied in anode and cathode gap and the workpiece is dissolved in electrolyte by anodic dissolution. The electrolyte chosen for this kind of operation is generally a weak acidic solution [64]. There are varieties of drilling processes of micro-ECM for machining of titanium alloy, for example, electrochemical drilling (ECD) and acid-based ECM drilling processes: shaped tube electrolytic machining (STEM), capillary drilling (CD), electro-stream drilling (ESD), and jet electrolytic drilling (JED). The acid-based electrochemical micro-drilling process has advantages over other processes. No residual stress is present and it can deliver excellent surface finish. Tool wear, burr formation or distortion of the holes, neither of these is an issue in this process which is fairly common in most of the other conventional and nonconventional machining processes. Also, a large number of holes can be produced simultaneously with this method [62]. The schematic of different kind of electrochemical drilling processes is presented in Fig. 19 [62].

As mentioned earlier, titanium and its alloys are widely used in biomedical industries due to their biocompatibility, high strength-to-weight ratio, and excellent corrosion resistance. While dealing with implants, surface topography of nano to millimeter scale is important, especially in load bearing implants. Surface features of hundred-micron level are necessary for mechanical interlocking of bone tissue and better osteointegration. So micromachined titanium surfaces and their biological characteristics with considerable surface topographies has become a topic of great interest nowadays [65, 66]. Wet chemical etching, reactive ion etching, and laser patterning are some of the most popular methods for conducting micro-fabrication on titanium surface. Each of these comes with certain advantages and disadvantages. Wet chemical etching has certain level of safety and disposal issue as it uses toxic chemical etchant (in case of titanium, it is mainly hydrofluoric acid (HF) which is dangerously toxic). Additionally, it is really difficult to maintain the etch rate and shape evolution in this method. Reactive ion etching, being able to remove material with high precision is a popular choice in semiconductor industry. But very low etching rate can be achieved in this compared to its high equipment cost. Moreover, redeposition in titanium surface causes acute surface chemical heterogeneity problem in this method. In laser micromachining, metal melting during the machining process occurs which might cause a chemical property change and also it leaves residue stress on implant surface [67].

Electrochemical micromachining, which can remove material selectively by electrochemical reaction at anode, is also a popular choice of producing micropattern on titanium surface. Recent developments in this field show a lot of advantages of using electrochemical micromachining (EMM). Surface properties are not in threat to be changed in this process as it deals with atom by atom removal of material. Unlike laser micromachining or dry etching process, generation of residual stress is not an issue here. Again, in EMM noncorrosive, not-toxic and environment-friendly chemicals are used. Also, the machining rate can be adjusted by controlling electrical current which gives the whole process a smooth flexibility [67]. Through mask electrochemical micromachining (TMEMM) is one special kind which can create micropatterns on photoresist-coated titanium substrates using photolithography. Titanium then gets dissolved from the unprotected area [68–72]. However, one disadvantage of TMEMM is its inability to produce micro-features with high aspect ratio in titanium alloys (Fig. 13).

Jet electrochemical micromachining (Jet-EMM) is a newly developed method based on localized electrochemical reaction to produce micro-holes with high aspect ratio on titanium surface [73–76]. The schematic of Jet-EMM system is shown in Fig. 14a. Ti-6Al-4V cylinder of length 50 mm and diameter 12 mm were used as material and 5 mol/L NaBr aqueous solution was used for titanium etching. SEM was used to investigate three-dimensional geometric patterns and a surface profiler was used to measure the dimensions of micropatterns. Electrolyte jet flow is generated through a metal nozzle. A pressurized tank with compressed air is used for this purpose. Controlling the diameter of the nozzle, the jet flow along the titanium surface can be controlled. Constant voltage is maintained between the nozzle and the specimen through a high voltage power amplifier. The electrolyte starts acting as a cathode and the titanium workpiece as an anode and electrochemical reactions start to take place. An array of micro-holes machined on the Ti-6Al-4V by jet-EMM process is shown in Fig. 14b. Figure 15 shows the SEM images of a single micro-hole and 4×4 arrays of micro-holes machined on the Ti-6Al-4V using jet-EMM process. It was found that high aspect ratio burr-free micro-holes can be machined on difficult-to-cut material like titanium alloys using jet-EMM process.

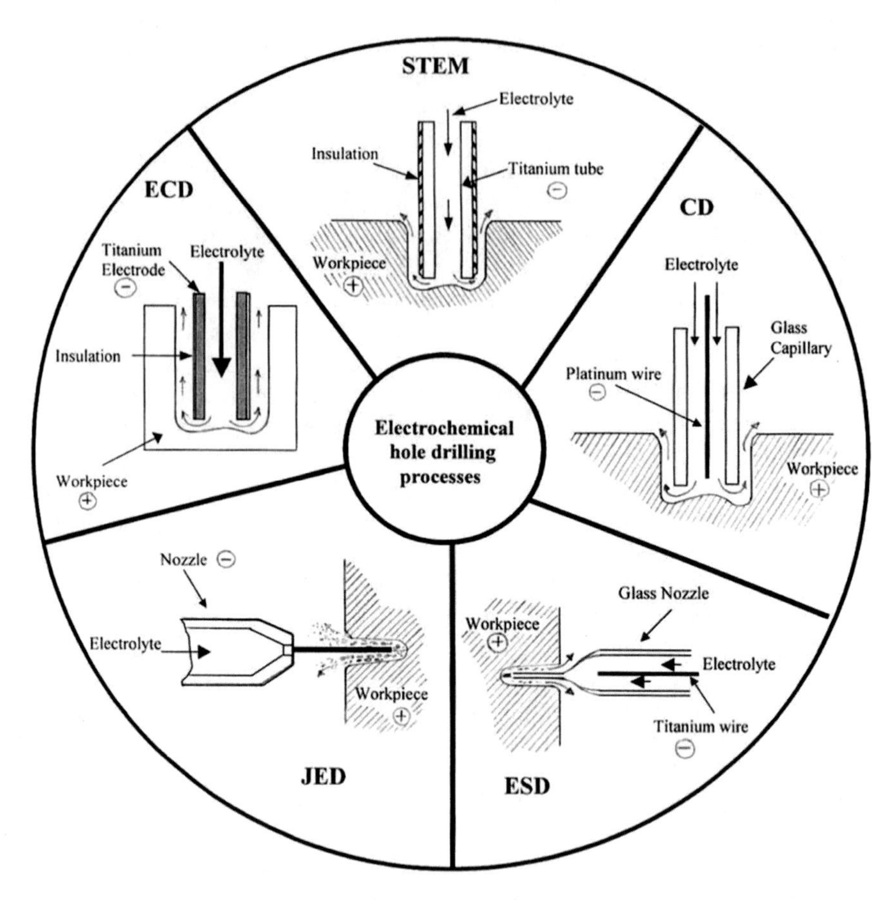

Fig. 13 Schematic diagrams of different ECM hole drilling process for machining difficult-to-cut materials like titanium alloys [62]

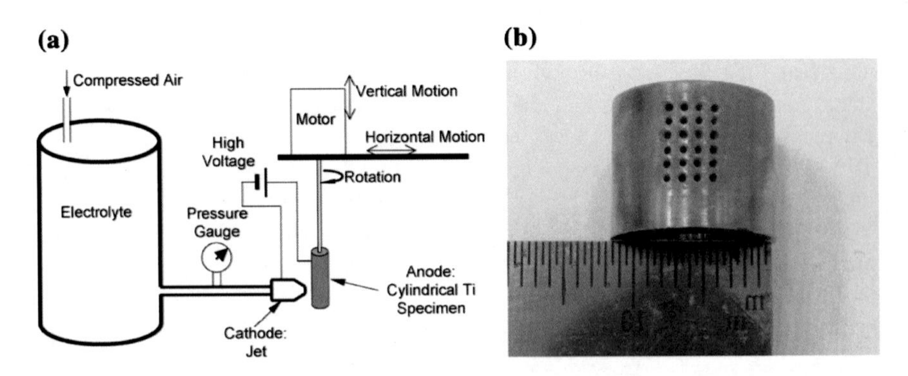

Fig. 14 **a** Schematic of Jet-EMM, and **b** micro-holes on Ti-6Al-4V by Jet-EMM [67]

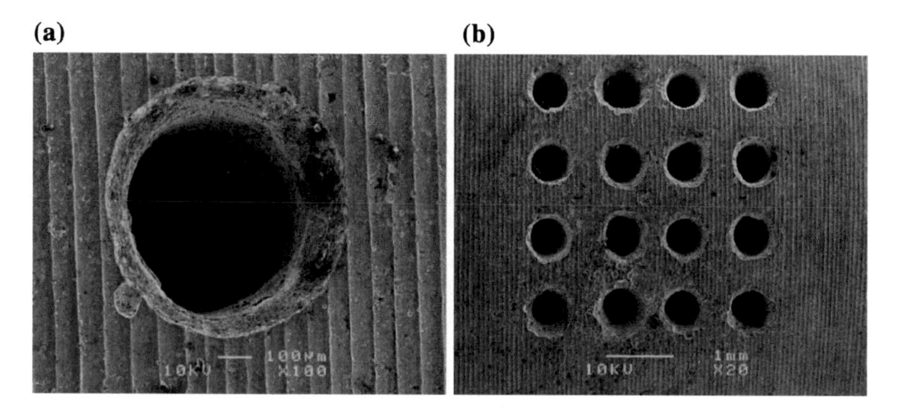

Fig. 15 SEM micrographs of **a** single hole and **b** 4 × 4 arrays of micro-holes on Ti-6Al-4V cylinder machined by Jet-EMM process [67]

3.3 *Laser Micromachining of Titanium Alloys*

Laser micromachining has become a popular nonconventional micromachining process for machining difficult-to-cut materials, due to its noncontact nature of material removal. It is widely used in many manufacturing industries, such as biomedical, electronics, and marine industries. Micro-drilling by laser has earned popularity in different aspects and now other laser micromachining techniques are being introduced by researchers. Besides laser micromachining, laser surface texturing (LST), and laser shock peening (LSP) are two of the newly developed methods for surface processing of hard and difficult-to-cut materials, which are being used successfully in different applications.

LST and LSP are found to be two useful laser micromachining process for machining and surface processing of titanium alloys. It has been found that working as a fluid reservoir micro-dent can promote better lifetime for mechanical parts as the lubricant gets trapped on those micro-dents and works as micro-bearings sustaining the load. It also helps in reducing the surface wear. While it is flooded enough, micro-dents can also be used as a temporary storeroom for wear debris [77]. It was found that micro-dent can improve surface lifetime ten times of its previous value. In rolling and sliding contact, it also helps in minimizing friction. There are a good number of manufacturing technique out there for fabricating dent arrays which can be categorized into mechanical, lithographic, energy beam, coating techniques, etc. Micro-indentation is pretty popular among micro-dent fabrication technique where the dent geometry follows the indenter tip geometry. It was found that indentation-produced dent can suppress pitting and improve scuffing resistance on surfaces in rolling–sliding contact [78]. Another popular method out there is micro–drilling, but here the cutting technique affects the surface integrity and small chips produced during the cutting are difficult to flush away from the micro-holes [79].

For all these drawbacks that come with the existing conventional micromachining processes, laser surface texturing (LST), a new method of fabricating dent arrays come into play, especially when dealing with difficult-to-machine material like titanium alloys. High-energy laser pulses are used in this process to melt and vaporize metal in machining zone. While creating dent array with LST, depth, diameter, and dimple density are among the most important parameters that should be considered. It was found that LST created dents promotes less tool wear and long fatigue life by improving surface interaction [80, 81]. The main disadvantages of LST are at high temperatures while ablation occurs, along with melting and cracking, change of surface microstructure takes place which can hugely affect the fatigue life of the material [82]. So the necessity of a new method arises that will fabricate micro-dent without damaging any material properties.

Laser shock peening (LSP) is a well-known method which can protect material failure due to fatigue and wear by changing subsurface of a material inducing deep compressive residual stress. It is found that LSP can improve fatigue performance of titanium alloy [83]. Investigation has been done with 2024-T3Al and it was noticed that LSP improves fatigue strength of the titanium alloy compared to initial strength [84]. Laser shock peening was also performed on Ti-6Al-4V on microscale for fabricating micro-dents. It was found that both dent depth and diameter increases with increasing laser power. For strain hardening and compressive residual stress, hardness of peened surface also increases [77]. Figure 16a shows the schematic representation of the laser shock peening process on the metal surface [77]. The surface topography and profiles of three different dents fabricated by LSP process using power settings of 1, 2, and 3 W are shown in Fig. 16b, c.

For a biomechanical implant to be successful in long run, the main two factors are mechanical properties of the implant material and its adjusting or anchorage capability in the host tissue. It is more important to consider those two factors while dealing with high load bearing implants. The chosen material should be biocompatible enough to deal with the hindrance of fatigue and should be able to tolerate the high ultimate stress. Also, the biomechanical interface should be such which promotes ideal transfer of load [85]. With the development of material science, there has been a boom in creating a different type of metal alloys which are serving their purpose near perfectly in so many cases. Titanium is no different in that. In some of the alloys, the yield and tensile strength are almost double than commercially pure titanium. Also, their resistance to fatigue fracture is better than the pure titanium and that is why they are more preferred in biomedical application [86, 87]. Now, the most popular alloy in implant industry, especially in hip and knee joint implants is grade V titanium alloy (90% Ti, 6% Al, and 4% V, depicted Ti-6Al-4V). Alteration of surface structure of the titanium implants can provide increased torque capacity and fatigue strength, which can promote biomechanical stability of implants inside the human body. Quite a few number of surface treatment methods are available in dentistry and orthopedics among which oxidation, blasting, and plasma spraying are commonly used [88].

Laser surface treatment is also a popular method of altering surface properties of the implant areas which are associated with higher bone formation to promote faster

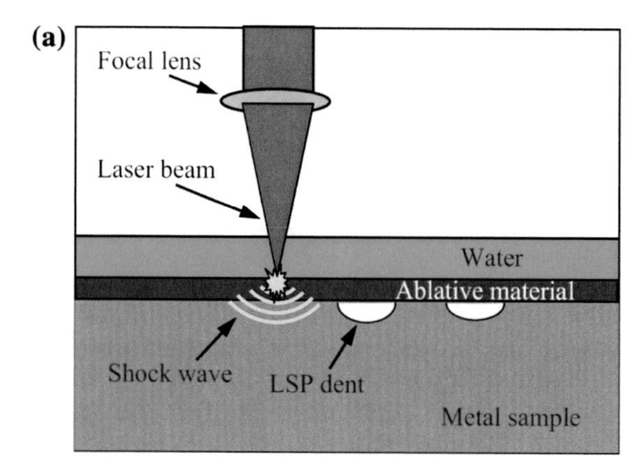

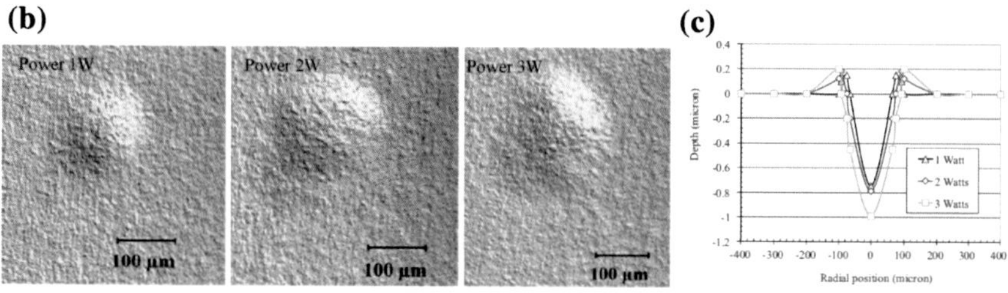

Fig. 16 **a** Schematic representation of dent fabrication on metal surface by laser shock peening method, **b** surface topography of three dents formed by LSP process using different power settings, and **c** profiles of the dents shown in (**b**) [77]

anchorage of biomechanical implant in human tissue. It can keep the rest of the surface as machined while changing the surface properties of the area required. One of the main advantages of laser treatment is that unlike other methods like oxidation, blasting, and plasma spraying, there is no need to remove or introduce any foreign material to change the surface property so the purity of material is sustained. However, change in microstructure and phase composition of the implant surface and subsurface might happen due to the heat associated with the laser treatment [85]. Moreover, the texture created on the surface by laser surface processing may promote cell adhesion and cell growth compared to plain machined surface, as shown in Fig. 17 [85]. An oxide layer is formed on the laser-processed surface of the implant material due to thermal nature of material removal in the laser processing. The oxide layer formed during laser surface processing of titanium alloy is mostly titanium oxide, which may act as a protective coating on the implant. However, the fracture and fatigue properties of the implant may be affected due to the formation of oxide layer. As shown in Fig. 18, a surface oxide layer as thick as 0.1–0.51 μm can be seen even in low-resolution images of laser-treated sample of titanium alloy. The EDS analysis confirms that apart from titanium and oxygen, aluminum is also present in the surface layer. In addition, needle form of structural appearance was found instead of round grains that were found on subsurface of

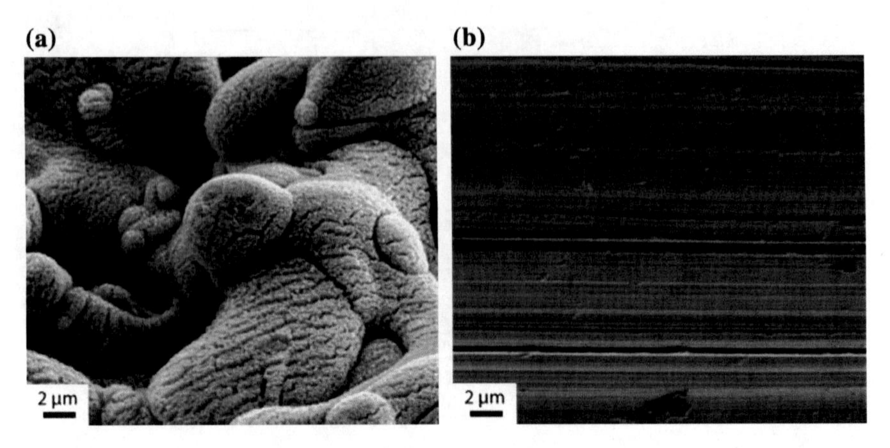

Fig. 17 SEM images of the surface topography of **a** laser-treated Ti-6Al-4V implant, **b** machined implant [85]

machined sample. Owing to the movement of melted material during the laser treatment, there is an increase in surface roughness and surface oxide thickness of laser-treated titanium alloy. As the temperature changes quite quickly, first raised and then decreased, reaction with the air at that time results in increased oxide thickness. At present, most of the laser-treated surface contains nano-structured titanium oxide but it is hard to identify because of the overlapping lattice parameters for rutile and anatase.

Apart from hip and knee joint implant, a lot of researches have been conducted in the field of dental implant and here as well, titanium is one of the most popular choice. Different kind of surface treatment and coating techniques have been carried out to improve the bone healing process of dental implant. There has been a newly developed method called direct laser fabrication (DLF), which can focus on metal powder microparticles in a laser beam to create the desired geometry. This is conducted through a computer 3D model and can be used to shape complex geometry solids. This method enables layer-by-layer development of the object which involves a fusion of titanium microparticles by computer-guided laser beam. This amazing method makes it possible to control the gradient of porosity in each layer of investigation. The porous surface promotes better bone healing process apart from better load adaptation and distribution [89]. As shown in Fig. 19, it is found that the highly mineralized bone matrix fills up all the gap among the peaks of the material surface. Inside the concavities and irregular grooves of the specimen, the bone matrix was noticed and it was closely connected with the surface and it was hardly displaced even by any traumatic action [89].

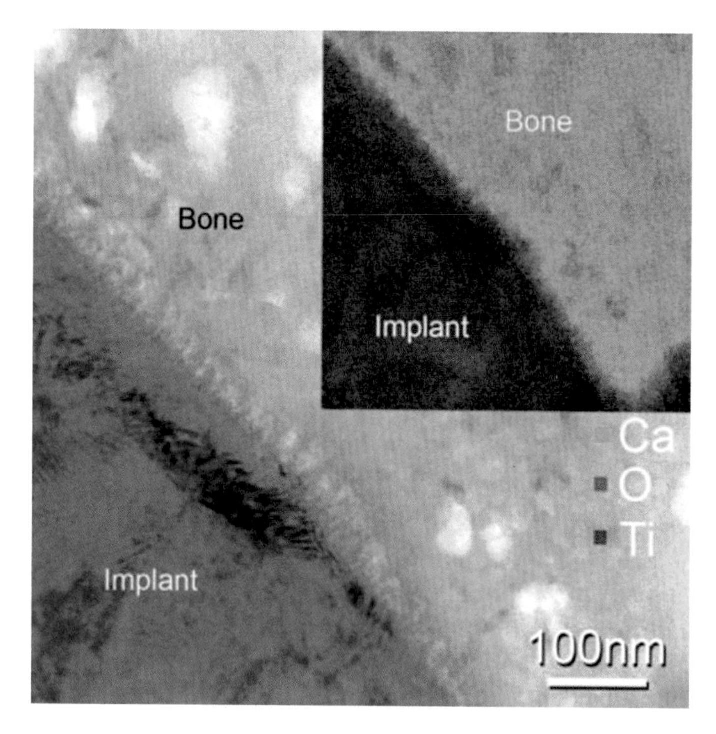

Fig. 18 The TEM and FETEM micrographs of the laser-processed implant surface showing a thin oxide layer at the interface between bone and implant [85]

Fig. 19 **a** Mineralized bone matrix on implant surface on electronic microphotograph, **b** attachment of bone matrix to implant surface on higher magnification [89]

3.4 *Abrasive Water-Jet Micromachining of Titanium Alloys*

Abrasive water-jet (AWJ) machining has been one of the fastest growing non-conventional machining process in recent years [90]. It has certain manufacturing advantages that very few machining techniques can provide. Though in most cases, the use of abrasive water-jet (AWJ) nozzle has been confined to machine features greater than 200 μm, recent research and advancement made it possible to machine in a much lower scale [91]. It is a newly developed excellent process to machine hard and difficult-to-cut materials like titanium and its alloys.

The AWJ micromachining process is material independent, that means it can cut or machine material according to their machinability with a single tool while in most conventional process different tool settings are needed for different materials. It can machine a huge range of versatile materials. Engineering polymers like—acrylic, Teflon, plastics, rubber, wood, and others, ductile and brittle materials like hardened steel, titanium alloy, ceramics, glass, stone, and many others, and also delicate materials like silicon, laminates, and composites, etc.—all can be machined by abrasive water-jet. It can cut materials of a wide range of thickness almost without any taper. As it is a cold cutting process, the disadvantages associated with cutting or melting creating heat-affected zone is not an issue here [91]. In this process, very little amount of force is exerted on the workpiece, which makes it possible to machine large aspect-ratio slots and ribs on thin shims [92]. With a single tool of abrasive jet, almost all modes of machining like cutting, turning, trimming, milling, beveling, etc., can be done.

In an AWJM system, the flow controller serves a lot of purpose. It receives pressurized water from the pumping unit. The flow controller directs all the water from the pump to the cutting nozzle, and simultaneously to displace the water from the top of the vessel to flow controller, it directs a small amount of water to abrasive storage vessel. Depending on the abrasive mixture concentration in the storage vessel and the required abrasive concentration at nozzle, the flow pump directs water partially from the pump to the top of abrasive storage vessel when abrasive is turned on. It helps abrasive mixture to be diluted by the other partial water that does not go to storage by displacing the mixture from the storage. By venting compressed water from the top of abrasive storage vessel, flow controller depressurize the system [90]. Suspension abrasive water-jet is one of the most efficient ways to do the process which is done by passing a pressurized suspension of an abrasive particle in water through a ceramic or diamond cutting nozzle. In this method, five times more cut surface area per minute can be achieved than a high-velocity water-jet which uses entraining abrasive particles carried in air [93].

Study has been done on abrasive water-jet machining of Ti-6Al-4V with different abrasive particles, such as white and brown aluminum oxide, garnet, glass beads, and steel shots. Most of the studies focused on investigating the effect of machining parameters on different performance characteristics. It was found that acting as a rigid indenter, particle hardness increases material removal rate and surface roughness. However, particle hardness along with shape factor do not

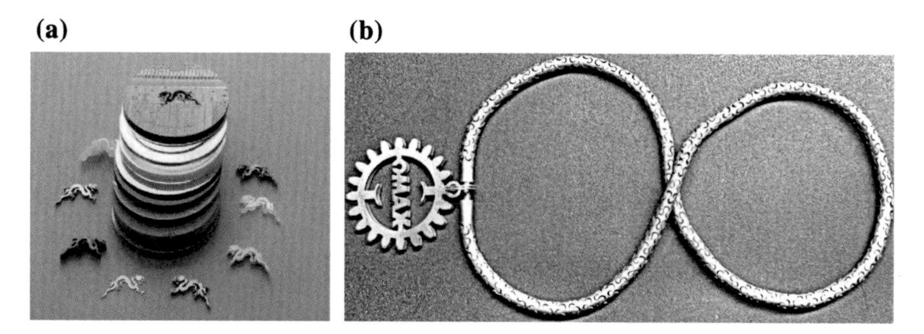

Fig. 20 **a** Machining of micro-features on titanium alloy cut by 40–60 μm jets [90], and **b** interlocking link of titanium tube made by abrasive water-jet machining [91]

influence surface waviness to a noticeable extent [94]. There have been studies focusing on fabrication of complex micro-features on titanium alloy using abrasive water-jet machining. Figure 20a shows complex 3D micro-features machined on titanium alloy and other different materials using 40–60 μm jets [90]. Figure 20b shows a complex interlocking link of titanium tube that was machined by abrasive water-jet machining [91].

4 Conclusions and Future Outlook

A comprehensive study on micromachining of titanium and its alloys have been presented in this book chapter. Both conventional and nonconventional micromachining processes have been covered and discussed thoroughly by describing the process in brief, key parameters, performance criterions, challenges, and so on. A good number of research on the respective field has been included to present an overview of each micromachining process for machining of titanium alloys. The conclusions that can be drawn from this book chapter are given below:

- In most of the conventional micromachining processes (micro-milling, micro-drilling, micro-turning), specially in micro-milling, burr formation is the main problem as in most cases parameters that provide minimum burr formation delivers comparably higher surface roughness. So finding an optimized setting of machining parameters that works for both minimizing the burr formation and improving the surface finish is a challenging task in conventional micromachining of titanium alloys.
- The necessity of producing micro-holes in difficult-to-cut materials like titanium alloys are increasing in industries and that is why micro-drilling has become one of the prime manufacturing process used extensively for machining of titanium alloys. Though there are a lot of micro-drill tools available, in conventional micromachining it is difficult to maintain accuracy at the small scale. As a result, micro-drilling by laser, micro-EDM, micro-ECM, and AWJ micromachining is

becoming more popular in recent years than the conventional mechanical micro-drilling process, especially for machining of titanium alloys.

- High cutting forces and tool temperature are two major challenges in mechanical micromachining of titanium alloys. Among various innovative solutions, micro-grooved tool in turning was found to decrease the cutting forces and cutting tool tip temperature to a great extent for micromachining of titanium alloys.
- Laser surface texturing and laser shock peening are two newly developed methods of micromachining and surface texturing, that could also improve the mechanical properties of the machined part. The fatigue life of titanium parts can be improved by changing subsurface of the material inducing residual stress by means of laser shock peening.
- Electric discharge machining (EDM) aided with ultrasonic vibration and ultrasonic vibration coupled with planetary movement are found to be effective to improve desired performance criterions while micromachining titanium alloys. Use of powder-mixed dielectric in micro-EDM also delivers better surface topography and minimizes recast layer issues for micro-EDM of titanium alloys.
- Electrochemical micromachining (EMM) and abrasive water-jet (AWJ) micromachining are gaining popularity for machining difficult-t-cut materials like titanium alloys, as in both cases challenges, i.e., high mechanical stress, tool wear, creation of heat-affected zone, etc., are minimized to a great extent.

There are some unsolved issued while micromachining of titanium alloys that should be considered by future researchers.

- In conventional micromachining processes, limiting burr formation and reducing surface roughness simultaneously is a real challenge as they work for opposing parameter settings. Removing the burr with another machining process can be an option in this case. Tool wear is also prominent in all major conventional micromachining process which needs to be improved. Close monitoring of tool path while machining, improvisation on tool geometry, and using different kind of tool coatings can minimize the problems associated with tool wear in conventional micromachining of titanium alloys.
- Very minimum amount of research has been progressed on hybrid micromachining, combining two or more micromachining processes in a single setup for achieving the final goal. Sometimes one challenge associated with one micromachining process can be solved by another process, and more study on this topic should be conducted to overcome the remaining challenges in micromachining of titanium alloys.
- As most of the micromachining processes deal with mechanical and thermal energies for material removal, change in microstructure after machining process can be an important consideration, and minimal research work has been done on this topic while micromachining of titanium alloys. This is one important topic that needs to be considered by future researchers.

- Nonconventional micromachining processes, such as laser micromachining and micro-EDM have mainly two issues: poor surface roughness and heat-affected zone. These two issues, though addressed before, are still to be solved properly. Introduction of hybrid micromachining processes combining another process (es) with laser or micro-EDM could minimize the negative effect on surface roughness and recast layer formed during laser micromachining and micro-EDM of titanium alloys.

References

1. B.B. Pradhan, M. Masanta, B.R. Sarkar, B. Bhattacharyya, Investigation of electro-discharge micro-machining of titanium super alloy. Int. J. Adv. Manuf. Technol. **41**(11), 1094–1106 (2009)
2. Titanium Grades Information—Properties and Applications for All Titanium Alloys; Pure Grades—Supraalloys.com, http://www.supraalloys.com/titanium-grades.php. Accessed 24 Dec 2017
3. M. Rahman, S. Wong, A.R. Zareena, Machinability of titanium alloys. JSME Int. J. Ser. C Mech. Syst. Mach. Elem. Manuf. **46**(1), 107–115 (2003)
4. T. Thepsonthi, T. Özel, Multi-objective process optimization for micro-end milling of Ti-6Al-4V titanium alloy. Int. J. Adv. Manuf. Technol. **63**(9–12), 903–914 (2012)
5. T. Thepsonthi, T. Özel, 3-D finite element process simulation of micro-end milling Ti-6Al-4V titanium alloy: experimental validations on chip flow and tool wear. J. Mat. Process. Technol. **221**, 128–145 (2015)
6. D. Umbrello, Finite element simulation of conventional and high speed machining of Ti6Al4V alloy. J. Mat. Process. Technol. **196**(1), 79–87 (2008)
7. M. Calamaz, D. Coupard, F. Girot, A new material model for 2D Numerical simulation of serrated chip formation when machining titanium alloy Ti-6Al-4V. Int. J. Mach. Tools Manuf. **48**(3), 275–288 (2008)
8. M. Sima, T. Özel, Modified material constitutive models for serrated chip formation simulations and experimental validation in machining of titanium alloy Ti-6Al-4V. Int. J. Mach. Tools Manuf. **50**(11), 943–960 (2010)
9. M. Hasan, J. Zhao, Z. Jiang, A review of modern advancements in micro drilling techniques. J. Manuf. Process. **29**, 343–375 (2017)
10. D.-W. Kim, M.-W. Cho, T.-I. Seo, E.-S. Lee, Application of design of experiment method for thrust force minimization in step-feed micro drilling. Sensors **8**(1), 211–221 (2008)
11. P. Yongchen, T. Qingchang, Y. Zhaojun, A study of dynamic stresses in micro-drills under high-speed machining. Int. J. Mach. Tools Manuf. **46**(14), 1892–1900 (2006)
12. B.H. Li, Z. Liu, H. Wang, Z.S. Zhong, Study on the stainless steel 1Cr18Ni9Ti micro-hole drilling experiment, in *Applied Mechanics and Materials* (Trans Tech Publ, 2014), pp. 43–46
13. L.A. Kudla, Deformations and strength of miniature drills. Proc. Inst. Mech. Eng. Part B J. Eng. Manuf. **220**(3), 389–396 (2006)
14. B.X. Ma, M. Nomura, T. Kawashima, T. Shibata, Y. Murakami, M. Masuda, O. Horiuchi, Study on micro drilling—rotating bending fatigue of micro carbide drills, in *Key Engineering Materials* (Trans Tech Publ, 2009), pp. 45–48
15. Y. Gong, K.F. Ehmann, C. Lin, Analysis of dynamic characteristics of micro-drills. J. Mater. Process. Technol. **141**(1), 16–28 (2003)
16. H. Shi, J. Ning, H. Li, Performance analysis of micro drill bit with asymmetric helix groove. Circuit World **41**(1), 7–13 (2015)

17. H.-S. Yoon, R. Wu, T.-M. Lee, S.-H. Ahn, Geometric optimization of micro drills using Taguchi methods and response surface methodology. Int. J. Precis. Eng. Manuf. **12**(5), 871–875 (2011)
18. M. Figueroa, E. García, S. Muhl, S. Rodil, Preliminary tribological study and tool life of four commercial drills. Tribol. Trans. **57**(4), 581–588 (2014)
19. L. Zheng, C. Wang, L. Yang, Y. Song, L. Fu, Characteristics of chip formation in the micro-drilling of multi-material sheets. Int. J. Mach. Tools Manuf. **52**(1), 40–49 (2012)
20. Y. Zhuang, Optimizing the economic efficiency by micro-drill life improvement during deep-hole drilling in the 212-valve manufacturing process (2013)
21. J.S. Nam, P.-H. Lee, S.W. Lee, Experimental characterization of micro-drilling process using nanofluid minimum quantity lubrication. Int. J. Mach. Tools Manuf. **51**(7), 649–652 (2011)
22. J.S. Nam, D.H. Kim, H. Chung, S.W. Lee, Optimization of environmentally benign micro-drilling process with nanofluid minimum quantity lubrication using response surface methodology and genetic algorithm. J. Clean. Prod. **102**, 428–436 (2015)
23. L. Fu, S.-F. Ling, C.-H. Tseng, On-line breakage monitoring of small drills with input impedance of driving motor. Mech. Syst. Signal Process. **21**(1), 457–465 (2007)
24. H. Wang, X. Li, Q.M. Ju, Online monitoring system for micro-hole drilling based on rough set fuzzy control, in *Key Engineering Materials* (Trans Tech Publ, 2011) pp. 15–19
25. C.K. Huang, Y.S. Tarng, C.Y. Chiu, A.P. Huang, Investigation of machine vision assisted automatic resharpening process of micro-drills. J. Mater. Process. Technol. **209**(18), 5944–5954 (2009)
26. T.H. Chen, W.T. Chang, P.H. Shen, Y.S. Tarng, Examining the profile accuracy of grinding wheels used for microdrill fluting by an image-based contour matching method. Proc. Inst. Mech. Eng. Part B J. Eng. Manuf. **224**(6), 899–911 (2010)
27. L. Wang, L. Zheng, C. Yong Wang, S. Li, Y. Song, L. Zhang, P. Sun, Experimental study on micro-drills wear during high speed of drilling IC substrate. Circuit World, **40**(2), 61–70 (2014)
28. C.L. Chao, W.C. Lin, W.C. Chou, Y.T. Chen, K.J. Ma, C.W. Chao, Study on extending tool life of micro WC drills by various protective coatings, in *Applied Mechanics and Materials* (Trans Tech Publ, 2012) pp. 2121–2124
29. C. Miao, F. Qin, G. Sthur, K. Chou, R. Thompson, Integrated design and analysis of diamond-coated drills. Comput. Aided. Des. Appl. **6**(2), 195–205 (2009)
30. T. Szalay, K. Patra, B.Z. Farkas, experimental investigation of tool breakage in micro drilling of EN AW-5083 aluminium, in *Key Engineering Materials* (Trans Tech Publ, 2014), pp. 119–124
31. H.E. Sevil, S. Ozdemir, Prediction of microdrill breakage using rough sets. AI EDAM **25**(1), 15–23 (2011)
32. X. Yang, C. Richard Liu, Machining titanium and its alloys. Mach. Sci. Technol. **3**(1), 107–139 (1999)
33. E.O. Ezugwu, Z.M. Wang, Titanium alloys and their machinability—a review. J. Mater. Process. Technol. **68**(3), 262–274 (1997)
34. T. Masuzawa, H.K. Tönshoff, Three-dimensional micromachining by machine tools. CIRP Ann. Technol. **46**(2), 621–628 (1997)
35. M.A. Rahman, M. Rahman, A.S. Kumar, H.S. Lim, A. Asad, Fabrication of miniature components using microturning, in *Proceedings of the Fifth International Conference on Mechanical Engineering* (Dhaka, 2003)
36. V. Senthilkumar, S. Muruganandam, State of the art of micro turning process. Int. J. Emerg. Technol. Adv. Eng. ISSN, 2250–2459 (2012)
37. H.S. Lim, A.S. Kumar, M. Rahman, Improvement of form accuracy in hybrid machining of microstructures. J. Electron. Mater. **31**(10), 1032–1038 (2002)
38. A.S. Patil, H.K. Dave, R. Balasubramaniam, K.P. Desai, H.K. Raval, Some preliminary metallurgical studies on grain size and density of work material used in micro turning operation. J. Miner. Mat. Charact. Eng. **9**(9), 845 (2010)

39. C.H.C. Haron, A. Ginting, H. Arshad, Performance of alloyed uncoated and CVD-coated carbide tools in dry milling of titanium alloy Ti-6242S. J. Mat. Process. Technol. **185**(1), 77–82 (2007)
40. T. Sugihara, T. Enomoto, Development of a cutting tool with a nano/micro-textured surface—improvement of anti-adhesive effect by considering the texture patterns. Precis. Eng. **33**(4), 425–429 (2009)
41. T. Obikawa, A. Kamio, H. Takaoka, A. Osada, Micro-texture at the coated tool face for high performance cutting. Int. J. Mach. Tools Manuf. **51**(12), 966–972 (2011)
42. J. Xie, M.J. Luo, K.K. Wu, L.F. Yang, D.H. Li, Experimental study on cutting temperature and cutting force in dry turning of titanium alloy using a non-coated micro-grooved tool. Int. J. Mach. Tools Manuf. **73**, 25–36 (2013)
43. X. Xu, Y. Yu, H. Huang, Mechanisms of abrasive wear in the grinding of titanium (TC4) and nickel (k417) alloys. Wear **255**(7), 1421–1426 (2003)
44. Z.Y. Yu, Y. Zhang, J. Li, J. Luan, F. Zhao, D. Guo, High aspect ratio micro-hole drilling aided with ultrasonic vibration and planetary movement of electrode by micro-EDM. CIRP Ann. Technol. **58**(1), 213–216 (2009)
45. Z.Y. Yu, K.P. Rajurkar, H. Shen, High aspect ratio and complex shaped blind micro holes by micro EDM. CIRP Ann. Technol. **51**(1), 359–362 (2002)
46. G. Kibria, B.B. Pradhan, B. Bhattacharyya, Experimentation and analysis into micro-hole machining in EDM on Ti-6al-4v alloy using boron carbide powder mixed de-Ionized water. Int. J. Mat. Manuf. Des. Acad. Res. J. 17–35 (2012)
47. Q.H. Zhang, R. Du, J.H. Zhang, Q.B. Zhang, An investigation of ultrasonic-assisted electrical discharge machining in gas. Int. J. Mach. Tools Manuf. **46**(12), 1582–1588 (2006)
48. B.H. Kim, C.N. Chu, Micro electrical discharge milling using deionized water as a dielectric fluid. J. Micromech. Microeng. **17**(5), 867 (2007)
49. G. Kibria, B.R. Sarkar, B.B. Pradhan, B. Bhattacharyya, Comparative study of different dielectrics for micro-EDM performance during microhole machining of Ti-6Al-4V alloy. Int. J. Adv. Manuf. Technol. **48**(5), 557–570 (2010)
50. M.L. Jeswani, Electrical discharge machining in distilled water. Wear **72**(1), 81–88 (1981)
51. H. Narumiya, EDM by powder suspended working fluid, in *Proceedings of 9th International Symposium on Electromachining (ISEM IX)* (1989)
52. N. Mohri, N. Saito, M. Higashi, N. Kinoshita, A new process of finish machining on free surface by EDM methods. CIRP Ann. Technol. **40**(1), 207–210 (1991)
53. Y.S. Wong, L.C. Lim, I. Rahuman, W.M. Tee, near-mirror-finish phenomenon in EDM using powder-mixed dielectric. J. Mater. Process. Technol. **79**(1), 30–40 (1998)
54. Y. Uno, A. Okada, T. Yamada, Y. Hayashi, Y. Tabuchi, Surface integrity in EDM of aluminum bronze with nickel powder mixed fluid. J. Japan Soc. Electr. Mach. Eng. **32**(70), 24–31 (1998)
55. W.S. Zhao, Q.G. Meng, Z.L. Wang, The application of research on powder mixed EDM in rough machining. J. Mater. Process. Technol. **129**(1), 30–33 (2002)
56. Y.-F. Tzeng, C.-Y. Lee, Effects of powder characteristics on electrodischarge machining efficiency. Int. J. Adv. Manuf. Technol. **17**(8), 586–592 (2001)
57. H.K. Kansal, S. Singh, P. Kumar, Technology and research developments in powder mixed electric discharge machining (PMEDM). J. Mater. Process. Technol. **184**(1), 32–41 (2007)
58. M.L. Jeswani, Effect of the addition of graphite powder to kerosene used as the dielectric fluid in electrical discharge machining. Wear **70**(2), 133–139 (1981)
59. A. Erden, S. Bilgin, Role of impurities in electric discharge machining, in *Proceedings of the Twenty-First International Machine Tool Design and Research Conference* (Springer, 1981), pp. 345–350
60. G.E. Baker, Hole drilling processes: experiences, applications, and selections, in *SME Non-Traditional Machining Symposium, Orlando, Florida, 3rd–5th, Amchem Company* (1991), pp. 6–12

61. J. Bannard, Fine hole drilling using electrochemical machining, in *Proceedings of the Nineteenth International Machine Tool Design and Research Conference* (Springer, 1979), pp. 503–510
62. M. Sen, H.S. Shan, A review of electrochemical macro-to micro-hole drilling processes. Int. J. Mach. Tools Manuf. **45**(2), 137–152 (2005)
63. J. Kozak, K.P. Rajurkar, R. Balkrishna, Study of electrochemical jet machining process. Trans. Soc. Mech. Eng. J. Manuf. Sci. Eng. **118**, 490–498 (1996)
64. H.S. Shan, *Advanced Manufacturing Methods* (New Delhi, 2004)
65. X. Lu, Y. Leng, Quantitative analysis of osteoblast behavior on microgrooved hydroxyapatite and titanium substrata. J. Biomed. Mater. Res. Part A **66**(3), 677–687 (2003)
66. N.A.F. Jaeger, D.M. Brunette, Production of microfabricated surfaces and their effects on cell behavior, in *Titanium in Medicine* (Springer, 2001), pp. 343–374
67. X. Lu, Y. Leng, Electrochemical micromachining of titanium surfaces for biomedical applications. J. Mater. Process. Technol. **169**(2), 173–178 (2005)
68. C. Madore, D. Landolt, Electrochemical micromachining of controlled topographies on titanium for biological applications. J. Micromech. Microeng. **7**(4), 270 (1997)
69. C. Madore, O. Piotrowski, D. Landolt, Through-mask electrochemical micromachining of titanium. J. Electrochem. Soc. **146**(7), 2526–2532 (1999)
70. P. Chauvy, C. Madore, D. Landolt, Electrochemical micromachining of titanium through a patterned oxide film. Electrochem. Solid-State Lett. **2**(3), 123–125 (1999)
71. P.-F. Chauvy, P. Hoffmann, D. Landolt, Electrochemical micromachining of titanium through a laser patterned oxide film. Electrochem. Solid-State Lett. **4**(5), C31–C34 (2001)
72. Y. Ferri, O. Piotrowski, P.F. Chauvy, C. Madore, D. Landolt, Two-level electrochemical micromachining of titanium for device fabrication. J. Micromech. Microeng. **11**(5), 522 (2001)
73. M. Datta, Micromachining by electrochemical dissolution. Micromach. Eng. Mater. 239 (2001)
74. M. Datta, L.T. Romankiw, D.R. Vigliotti, R.J. Von Gutfeld, Jet and laser-jet electrochemical micromachining of nickel and steel. J. Electrochem. Soc. **136**(8), 2251–2256 (1989)
75. Y. Ito, M. Tada, Electrochemical technology, ed. by N. Masuko, T. Osaka, Y. Ito (1996)
76. J. Andrews, Electrochemical machining method and apparatus (1974)
77. Y.B. Guo, R. Caslaru, Fabrication and characterization of micro dent arrays produced by laser shock peening on titanium Ti–6Al–4V surfaces. J. Mater. Process. Technol. **211**(4), 729–736 (2011)
78. T. Nakatsuji, A. Mori, The tribological effect of mechanically produced micro-dents by a micro diamond pyramid on medium carbon steel surfaces in rolling-sliding contact. Meccanica **36**(6), 663–674 (2001)
79. C.R. Friedrich, Micromechanical machining of high aspect ratio prototypes. Microsyst. Technol. **8**(4), 343–347 (2002)
80. I. Etsion, G. Halperin, A laser surface textured hydrostatic mechanical seal. Tribol. Trans. **45**(3), 430–434 (2002)
81. I. Etsion, State of the art in laser surface texturing. Trans. ASME-F-J. Tribol. **127**(1), 248 (2005)
82. I. Iordanova, V. Antonov, S. Gurkovsky, Changes of microstructure and mechanical properties of cold-rolled low carbon steel due to its surface treatment by Nd: glass pulsed laser. Surf. Coat. Technol. **153**(2), 267–275 (2002)
83. J.-M. Yang, Y.C. Her, N. Han, A. Clauer, Laser shock peening on fatigue behavior of 2024-T3 Al alloy with fastener holes and stopholes. Mater. Sci. Eng. A **298**(1), 296–299 (2001)
84. R.K. Nalla, I. Altenberger, U. Noster, G.Y. Liu, B. Scholtes, R.O. Ritchie, On the influence of mechanical surface treatments—deep rolling and laser shock peening—on the fatigue behavior of Ti–6Al–4V at ambient and elevated temperatures. Mater. Sci. Eng. A **355**(1), 216–230 (2003)

85. A. Palmquist, F. Lindberg, L. Emanuelsson, R. Brånemark, H. Engqvist, P. Thomsen, Biomechanical, histological, and ultrastructural analyses of laser micro-and nano-structured titanium alloy implants: a study in rabbit. J. Biomed. Mat. Res. Part A **92**(4), 1476–1486 (2010)
86. A.S. Guilherme, G.E.P. Henriques, R.A. Zavanelli, M.F. Mesquita, Surface roughness and fatigue performance of commercially pure titanium and Ti-6Al-4V alloy after different polishing protocols. J. Prosthet. Dent. **93**(4), 378–385 (2005)
87. C.-W. Lin, C.-P. Ju, J.-H.C. Lin, A comparison of the fatigue behavior of cast Ti–7.5 Mo with Cp titanium, Ti–6Al–4V and Ti–13Nb–13Zr alloys. Biomaterials **26**(16), 2899–2907 (2005)
88. B. Al-Nawas, H. Götz, Three-dimensional topographic and metrologic evaluation of dental implants by confocal laser scanning microscopy. Clin. Implant Dent. Relat. Res. **5**(3), 176–183 (2003)
89. C. Mangano, J.A. Shibli, F. Mangano, R. Sammons, A. Macchi, Dental implants from laser fusion of titanium microparticles: from research to clinical applications. J. osseointegr. **1**(1), 2–14 (2009)
90. D.S. Miller, Micromachining with abrasive waterjets. J. Mat. Process. Technol. **149**(1), 37–42 (2004)
91. H.T. Liu, E. Schubert, D. McNiel, μAWJ technology for meso-micro machining, in *Proceedings of 2011 WJTA-IMCA Conference and Exposition* (2011), pp. 19–21
92. H.T. Liu, Y. Hovanski, D.D. Caldwell, R.E. Williford, *Low-Cost Manufacturing of Flow Channels with Multi-Nozzle Abrasive-Waterjets: A Feasibility Investigation* (Pacific Northwest National Laboratory (PNNL), Richland, WA (US), 2008)
93. R. Kovacevic, M. Hashish, R. Mohan, M. Ramulu, T.J. Kim, E.S. Geskin, State of the art of research and development in abrasive waterjet machining. J. Manuf. Sci. Eng. **119**(4B), 776–785 (1997)
94. G. Fowler, I.R. Pashby, P.H. Shipway, The effect of particle hardness and shape when abrasive water jet milling titanium alloy Ti6Al4V. Wear **266**(7), 613–620 (2009)

Ultra-precision Diamond Turning Process

Vinod Mishra, Harry Garg, Vinod Karar and Gufran S. Khan

1 Introduction

Surfaces with high finish and profile quality have gained popularity not only in high-end applications, i.e., aerospace, lithography system, biomedical engineering but also in common industrial applications such as CD/DVD lenses and metal mirrors [1, 2]. Deterministic, localized and micron to submicron material removal is required to achieve the nanometer level surface quality, which otherwise is not possible by traditional machining methods. Advancements in manufacturing technologies make it possible to fabricate the parts with critical surface finish, shape, and dimensional requirements. The ultra-precision machining is one of the mature technologies and it is possible to develop the surfaces of stringent profile requirements. The ultra-precision machining technology was first developed in the late seventeenth century for precision machining of mechanical components for aesthetics purpose and to meet the close dimensional tolerances [3, 4]. Later, due to advancements in manufacturing automation and control systems, the diamond turning process becomes more reliable for processing of components of high-quality surfaces of ductile and brittle materials, e.g., metals, polymers and crystals. However, it is still challenging to meet the increasing demands of the modern applications in terms of miniature size, nano-metric surface finish, submicron dimensional, and profile accuracy. It is crucial to optimize each stage of the process to meet these challenges. The quality of fabricated surface by ultra-precision machining is influenced by a number of input parameters, broadly

V. Mishra (✉) · H. Garg · V. Karar
CSIR-Central Scientific Instruments Organization, Chandigarh, India
e-mail: vnd.mshr@gmail.com

V. Mishra · G. S. Khan (✉)
Indian Institute of Technology, Instrument Design and Development Center, Delhi, India
e-mail: gufranskhan@iddc.iitd.ac.in

K. Kumar et al. (eds.), *Micro and Nano Machining of Engineering Materials*, Materials Forming, Machining and Tribology,
https://doi.org/10.1007/978-3-319-99900-5_4

categorized into two groups, viz., controllable and uncontrollable parameters. Understanding of both the controllable and the uncontrollable parameters and their relationship with the developed surface is required to achieve the desired surface quality.

Two-axis configuration of ultra-precision machining is mostly used for rotationally symmetric surfaces. The fabrication process typically starts with generation of initial shape by a conventional machining or grinding with suitable tolerance for final machining followed by ultra-precision machining with optimized machining parameters. The selection of machining parameters mainly depend on the type of the material and the selected tool type. Not only optimized parameters and quality of the tool, but the accurate tool setting is also very important to achieve the required profile [5–7]. Improper selection of machining parameters and poor holding conditions lead to vibrations, hence, results in mid-spatial frequencies and poor surface finish [8, 9]. The generation of vibration depends on number of sources such as machine dynamics, material properties, tool overhang, and tool geometry. Tool wear is another dominant factor that affects the cutting forces, surface finish, and the profile [10–12]. The rate of tool wear varies with workpiece material. The crystals, e.g., Si, Ge, CaF_2, ZnS are hard and brittle in nature and are difficult to machine [13]. In order to achieve submicron form accuracy, the process has to be corrected in a closed loop control [14]. Suitable metrology feedback is essential to give information about the fabricated surface and further to modify the tool path for corrective machining.

Optical industry is one of the most benefited industries by ultra-precision machining technology. The ongoing technological advances in areas of optical instrumentations set demands for more complex surfaces, which includes non-rotationally symmetric surfaces and micro-structured surfaces. Compared to the conventional surfaces, freeform surfaces help to make the system more compact, light in weight and improve the performance of optical system [15, 16]. The freeform shape with stringent profile is not possible to achieve by two-axis ultra-precision machining. Slow tool servo (STS) and Fast tool servo (FTS) configurations of ultra-precision machining are most suitable fabrication techniques for freeform surfaces [17, 18]. Both the techniques are capable to generate the freeform surface if supported by an appropriate metrology and feedback mechanism [19]. Material removal mechanism is different in FTS and STS machining as compared to two-axis machining and it is still challenging to achieve the good surface quality as compared to two-axis configuration [20].

Although significant progress is reported in different domains of ultra-precision machining in the last three decades and almost all areas are touched by individual researchers but still very less literature is available to provide both basics, practical and advance approaches in systematic way. The aim of this chapter is to provide the basic understanding of the practical issues involved with ultra-precision machining. The chapter provides detailed information on various steps of ultra-precision machining process and related challenges. It covers basic steps of ultra-precision machining, mechanism of material removal and metrology.

2 Ultra-Precision Diamond Turning Configuration

The machine design is an important aspect of ultra-precision machining. High-quality parts of the best-suited materials ensure the precision and differentiate the ultra-precision machining from the conventional machining. Main parts of ultra-precision machining are spindle assembly, machine base, control system, tool setting system and motion feedback control system. The spindle assembly is designed to ensure the sufficient stiffness and to minimize the thermal and vibration effects. Aerostatic bearings are mostly used to provide the accurate high-speed rotation. Machine base is generally made of epoxy granite to provide the rigidity to the whole machining system [21]. Pneumatic-based damping is incorporated to damp the vibrations. Machine control system controls all the movements of the slides and spindle with high precision. It is a mode of communication in between machine and user and offers the nano-metric programming resolution. Typical ultra-precision diamond turning machines are available with feedback resolutions of 10–20 pm, axis straightness of 100–300 nm, spindle stiffness 100–200 N/μm and programming resolution of single nanometer [22, 23]. Ultra-precision machining process is highly sensitive to change in environmental conditions, therefore, machines are kept in temperature, humidity, and vibration controlled surroundings. Tool setting system is used to provide the cutting point coordinates and tool nose radius to the machine controller. Cutting point coordinates and tool nose radius are found either with the linear variable differential transformer (LVDT) sensor or by special purpose optical system. The machines are arranged in different configurations, depending on the application. The important configurations are summarized in Table 1.

3 Materials and Their Applications

The workpiece materials for ultra-precision diamond turning include nonferrous metals, crystals, and polymers. Materials such as aluminum, copper, electroless nickel show good machinability for ultra-precision machining. Several polymers and crystals are also suitable for ultra-precision machining and used for many optical applications. Commonly used and machinable polymers are polycarbonate, polystyrene, acrylic, and nylon. The crystals which can be processed through ultra-precision machining are germanium, zinc selenide, zinc sulphide, calcium fluoride and silicon, etc. [25]. The most common materials which are processed by ultra-precision machining are given in Table 2. The processing of these materials for generation of different types of surfaces and their application is shown in Fig. 1. Some materials are difficult to process due to their hardness and chemical composition, e.g., carbon content in ferrous materials can chemically react with

Table 1 Ultra-precision diamond turning configurations [18, 24]

Configuration	Description	Product type	Schematic
Two-axis (X-Z) 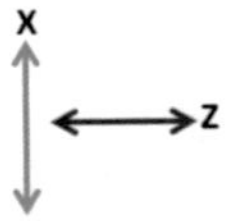	• Spindle on X-axis • Tool on Z-axis • Workpiece on X-axis	Rotationally symmetric surfaces, i.e., flat, spherical, aspheric, diffractive, etc.	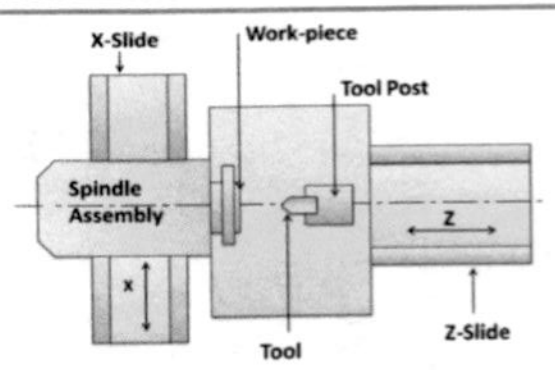
Two-axis ($R - \theta$) 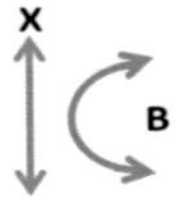	• Spindle on X-axis • Tool on B-axis • Workpiece on X-axis	Best suited for high-quality curved surfaces	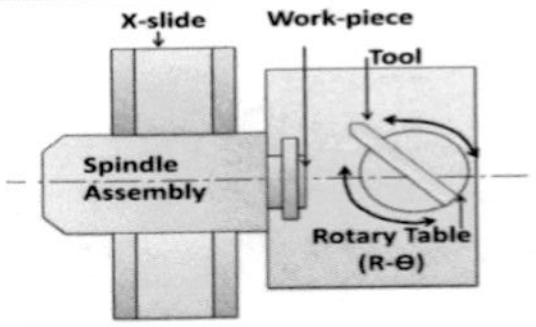
Slow tool servo (C-X-Z) 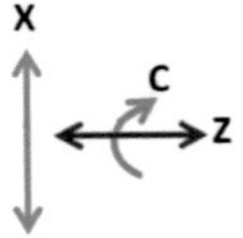	• Spindle on X-axis • Spindle rotation is controlled as a C-axis • Tool on Z-axis • Workpiece on X-axis	Rotationally symmetric, i.e., flat, spherical, aspheric, diffractive, non-rotationally symmetric, i.e., freeform, toric, off-axis	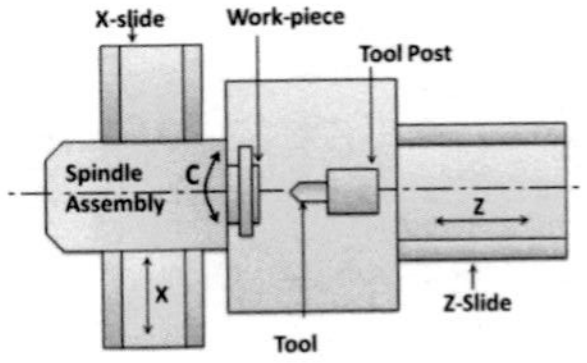

(continued)

Table 1 (continued)

Configuration	Description	Product type	Schematic
Fast Tool Servo (C-X-Z-W)	• Tool reciprocate at high frequency with the help of an attachment called fast tool servo and work as a W-axis • Spindle is on X-axis • Spindle rotation is controlled as a C-axis • Workpiece is on X-axis • Tool on Z-axis	Non-rotationally symmetric, i.e., Shallow freeform, toric, off-axis (with small sag only), Micro-structured surfaces, i.e., micro-lenslet array, microgrooves, micro functional features of different shapes	

Table 2 Important diamond turnable materials and their applications [30, 31]

Materials	Application	Application area
Metals		
Aluminum alloys	Mirrors, precision drums, memory disks, illumination optics	Automobile, aerospace, optics, nuclear, biomedical, defense
Copper	Mirrors, high power laser reflectors, molds	
Electroless nickel	Mold for micro-structured parts and optical lenses	
Titanium alloys	Biomedical implants, aerospace applications	
Brass	Mirrors, precision mechanical parts	
Beryllium copper	Molds, precision mechanical parts	
Crystals		
Silicon	Lenses for thermal imaging, IR microscopy, hyperspectral imaging, diffractive optics, axicon lenses	Defense, aerospace, electronics, MEMS
Germanium	IR optics, surveillance, rewritable DVDs, night vision, diffractive optics	
Zinc sulphide	Optical windows, lenses for visible to IR wavelength, resonators for CO_2 lasers	
Zinc selenide	Optical windows in FLIR, meniscus, cylindrical lenses	
Calcium fluoride	UV and IR thermal imaging, optical lenses, deep UV applications, spectroscopy	
Polymers		
Acrylic	Optical lenses, fresnel lenses, IOL's, lenslet array, LED lenses	Biomedical, ophthalmology, optics, instrumentation
Polycarbonate	Ophthalmic lenses, medical devices, projection optics	
Polysulfone	Fiber optics, medical instruments, food processing	

diamond and affect its life [26, 27]. Glass, due to its brittleness and titanium, due to its hardness are also difficult to process by ultra-precision diamond turning directly [28]. Figure 2 shows the images of actual fabricated components by ultra-precision machining.

4 Single Point Cutting Diamond Tools

Diamond has been used since decades for tooling and many more applications due to its unique properties, including hardness, coefficient of thermal expansion, chemical inertness, and thermal conductivity. Because of its wide range of

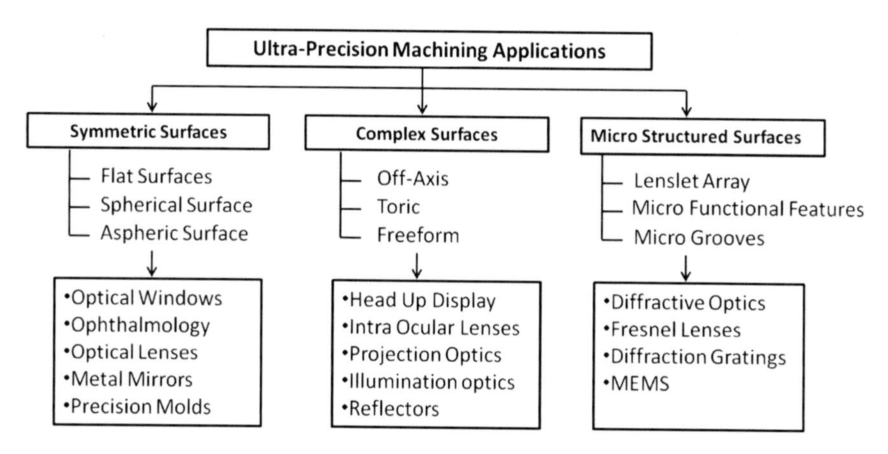

Fig. 1 Precision surfaces with their applications [2, 29]

properties, diamond is utilized as a tool material in ultra-precision machining [33]. Suitably cut diamonds are vacuum brazed into the metal shank. To reduce the vibrations and stiffness, tungsten, steel, molybdenum, and titanium are used as a tool shank material. A lattice structure of a diamond contains eight atoms, where each atom has four neighbor atom joined to it with covalent bond which makes it a high-density crystal. The high density of carbon atom and four neighbors for each atom in a single cell of crystal make diamond a super hard material [34]. Hardness of diamond varies directionally; so, choosing a right direction for preparation of cutting edge gives good results [35]. The toughness of the natural diamond is good as compared to the other gems but not comparatively good as other metals which are used for cutting. The poor toughness causes the chipping of the tool cutting edge; hence intermittent cutting is avoided in ultra-precision machining process. Strong bond between carbon atoms makes diamond an excellent heat conductor which is helpful to extract the heat from the cutting zone [34]. Like other cutting tools, diamond tools in ultra-precision machining are also specified by the value of front clearance angle, rake angle included angle, nose radius, cutting edge radius, and cutting arc quality [36, 37], as shown in Fig. 3.

The cutting edge quality of diamond tool is one of the main requirements for ultra-precision machining. The waviness or irregularities of cutting edge can replicate on the machined surface and limits the surface quality [38]. Waviness of cutting edge is usually controlled up to certain limits and if controlled in fraction of microns is considered as controlled waviness tool as shown in Fig. 4.

The quality of micro-structured components in terms of size and shape is mainly decided by quality and dimensions of the cutting tool [39]. Sharp cutting edge along with tip size of few micrometers is often required to generate the micro-features. The selection of tool tip dimensions and angles of cutting point depends on dimensions and the type of micro-feature required and machining configuration used for the fabrication [2]. Typical tools for ultra-precision machining of different surfaces are shown in Fig. 5.

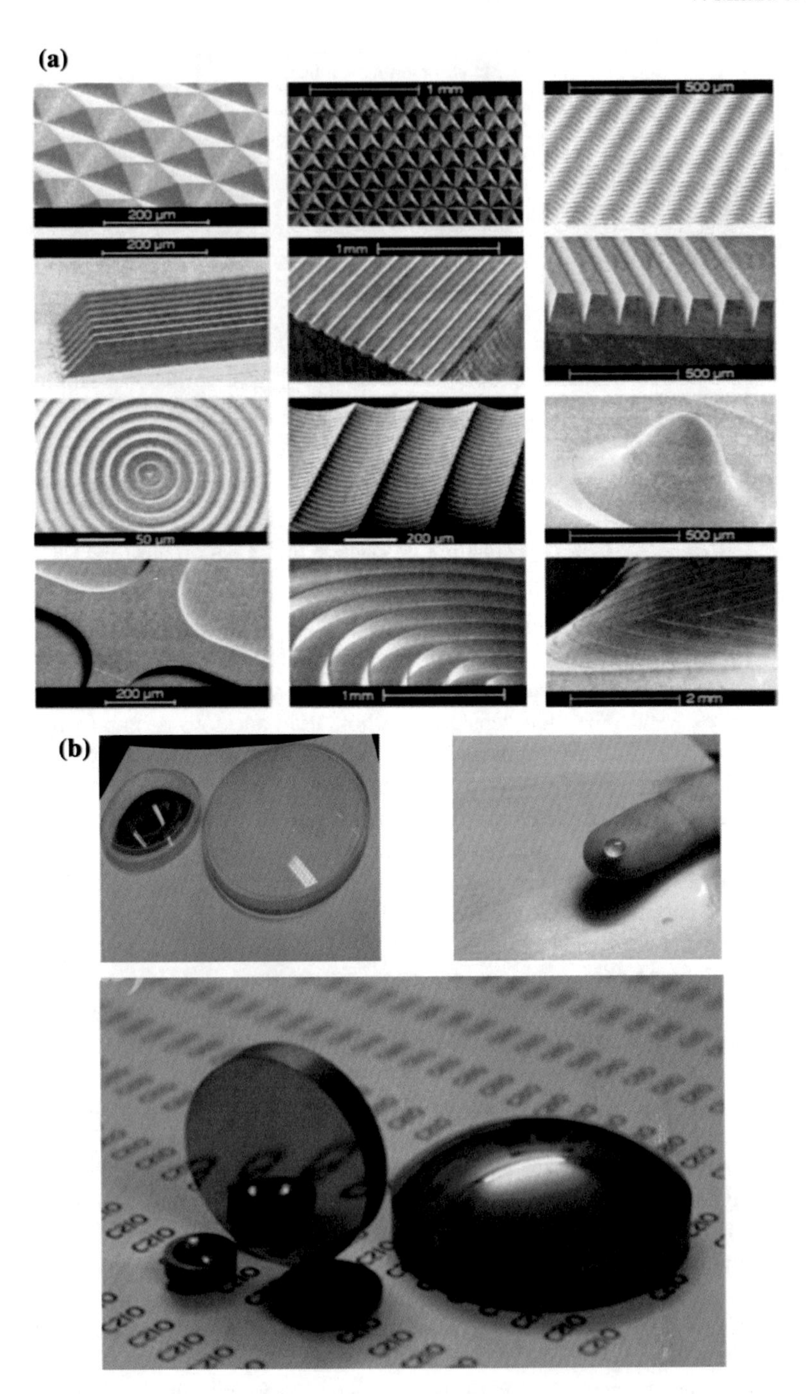

Fig. 2 Ultra-precision machining components **a** micro-structured components (Copyright SPIE), reproduced with permission from SPIE [32]. **b** Different optical components

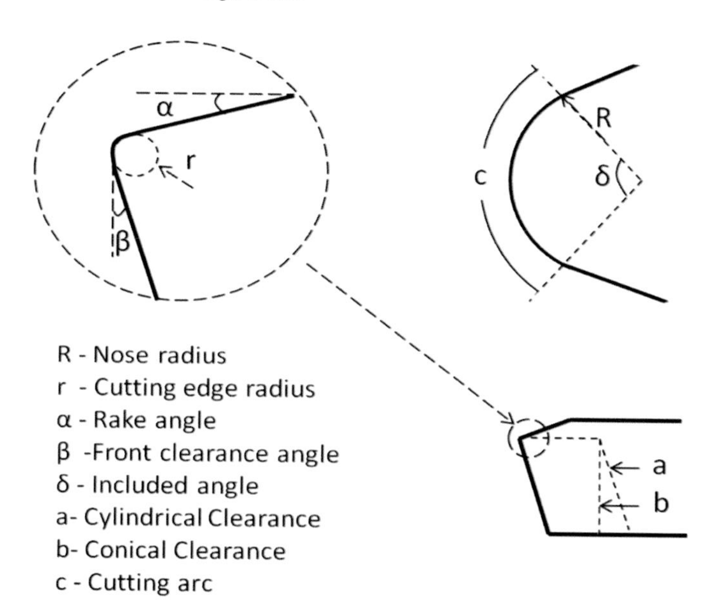

Fig. 3 Diamond tool specifications

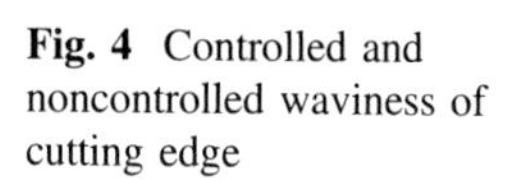

Fig. 4 Controlled and noncontrolled waviness of cutting edge

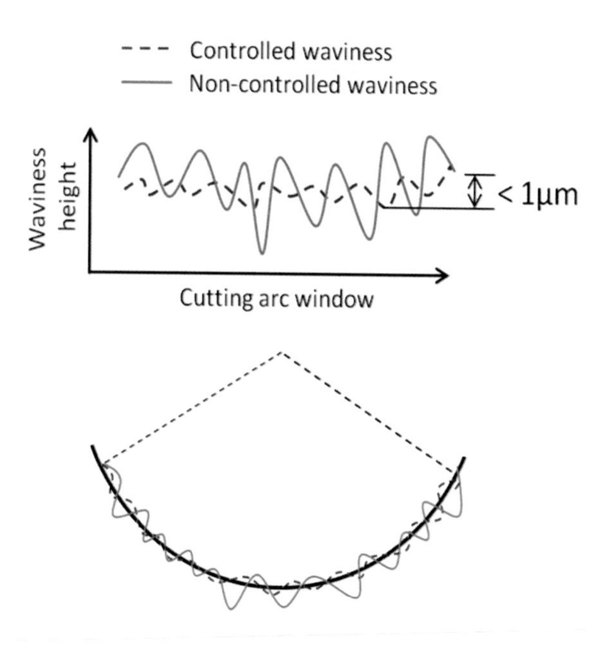

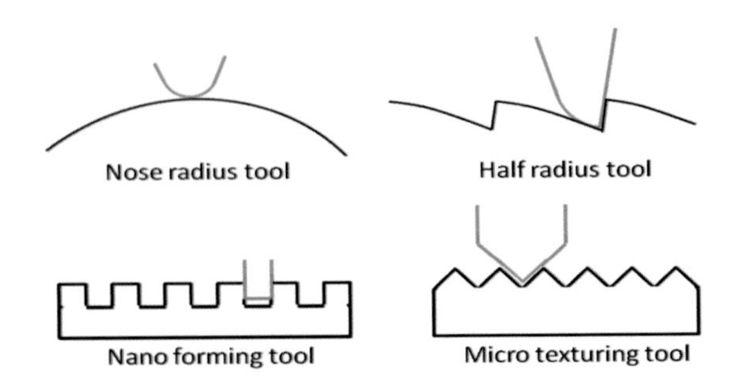

Fig. 5 Diamond tools for ultra-precision machining of different surfaces

5 Material Removal Mechanism

Material removal mechanism in any machining process is not only dependent upon the cutting tool and workpiece material but also dependent on selected process parameters and machine dynamics. Due to demand of high-quality surfaces, material removal is more sensitive to the above factors in ultra-precision machining. Conventionally, to achieve the nano-metric surface finish post-processing (lapping, honing, polishing) is required after machining. All these processes are force control process and not suitable for deterministic material removal. Due to nondeterministic nature, material removal is difficult to control in such processes [40]. On the other side, ultra-precision machining is displacement or path controlled process where the material is removed deterministically with sharp cutting point tool [25]. The other main difference in conventional machining and ultra-precision machining is the ratio of undeformed chip thickness to the tool cutting edge radius. Undeformed chip thickness is much larger than tool edge radius in conventional machining. Whereas, in ultra-precision machining, this ratio is very small [41]. Ultra-precision machining is generally associated with main three deformations, i.e., deformation due to contact between chip and tool rake face, deformation due to flank face rubbing and deformation because of material shear as shown in Fig. 6. Due to assumption of sharp cutting tool, most of the models for conventional machining are not valid for ultra-precision machining.

Ultra-precision machining is best suited for ductile polycrystalline materials and is explored more as compare to other materials. In ductile materials, chips are formed after significant plastic deformation followed by separation of material. Microstructure of material has an important role in removal mechanism due to size effects. The cutting energy required for separation depends upon defects density at the shear zone [42, 43]. Apart from above dissused deformations, ploughing is an important phenomenon for material removal in ultra-precision machining. When the undeformed chip thickness is of the level of cutting-edge radius, ploughing action takes place, rather than direct shearing as shown in Fig. 7 [44].

Brittle materials are difficult to process due to their hardness and crystalline nature. Material removal in case of brittle materials depends upon the

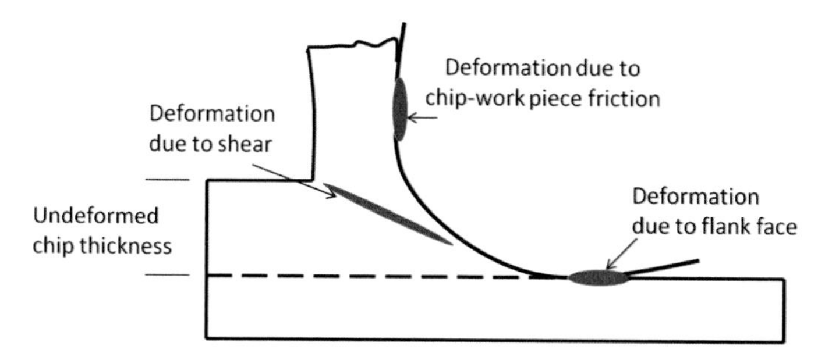

Fig. 6 Deformations during material removal process

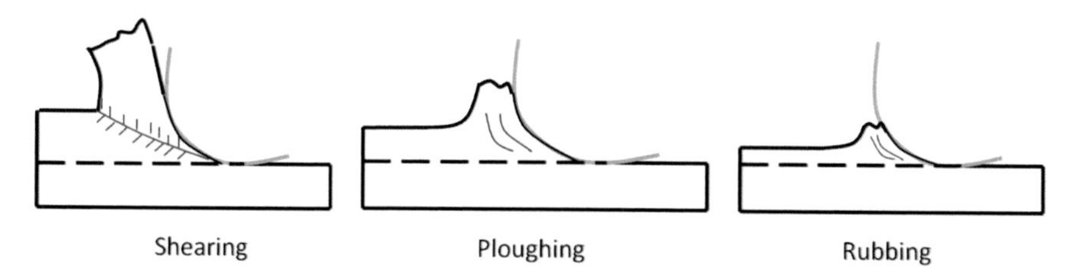

Fig. 7 Effect of cutting edge and depth of cut in material removal

crystallographic orientation which is responsible for variation of surface finish as well as cutting forces [45, 46]. However, at small depth of cuts, brittle materials behave like a ductile material. The depth of cut at which material removal is ductile is called critical depth of cut and the process is called ductile-regime machining. Depth of cut more than critical value leads to subsurface damage as shown in Fig. 8. Brittle to ductile transition also depends upon material properties and cutting forces [47].

Polymers are softer as compare to metals and crystals and having randomly or uniformly distributed polymer chains. Depending on polymer type and the structure, they behave differently during the machining deformation. At a particular temperature, called transition temperature T_g, solid polymers are converted into soft state [48]. At transition temperature, the material removal is converted from brittle to ductile. Other properties of polymers which affect the surface generation are their viscoelastic properties and thermal insulating nature [49].

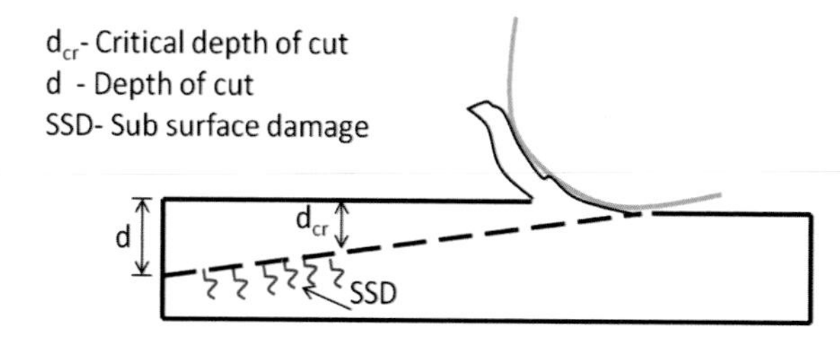

Fig. 8 Critical depth of cut for ductile-regime machining

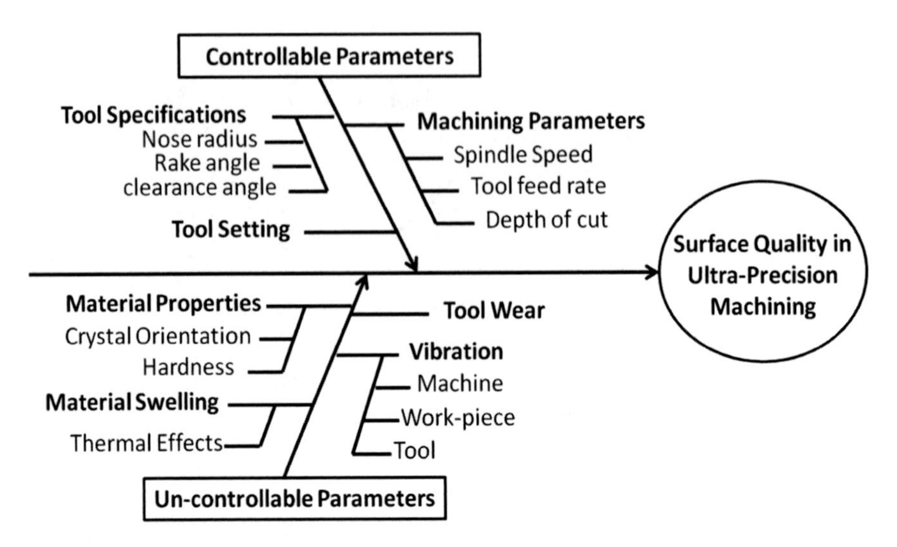

Fig. 9 Fish bone diagram for ultra-precision machining

6 Process Parameters for Ultra-Precision Machining

As discussed in previous sections, surface finish and profile accuracy are prime prerequisites of precision components and performance of ultra-precision machining decided by many factors which can be break into two groups, viz., controllable and uncontrollable. The spindle speed, tool feed, depth of cut, tool setting, and tool nose radius are the controllable parameters as operator can easily set their value. The parameters which are difficult to control and resulting from uncertain process conditions are called uncontrollable parameters. It includes vibration of tool and workpiece, material swelling, tool wear, and dynamic unbalance of machining system [50]. These factors deteriorate the surface quality collectively. Figure 9 lists the controllable and uncontrollable parameters.

6.1 Tool Setting

In ultra-precision machining, tool setting refers to the mounting and adjustment of the diamond tool tip with axis of spindle rotation. The close agreement between the programmed profile and the generated profile, without any center defect is the confirmation of tool setting. However, if the cutting tool tip is above, below or away from rotation axis, a tip or defect at the center of the workpiece is produced. For tool setting, machining is usually performed on test workpiece of curved profile. If the tool is not precisely located at the center of rotation axis, unwanted material is left at the center of the surface. The horizontal component of tool offset from the

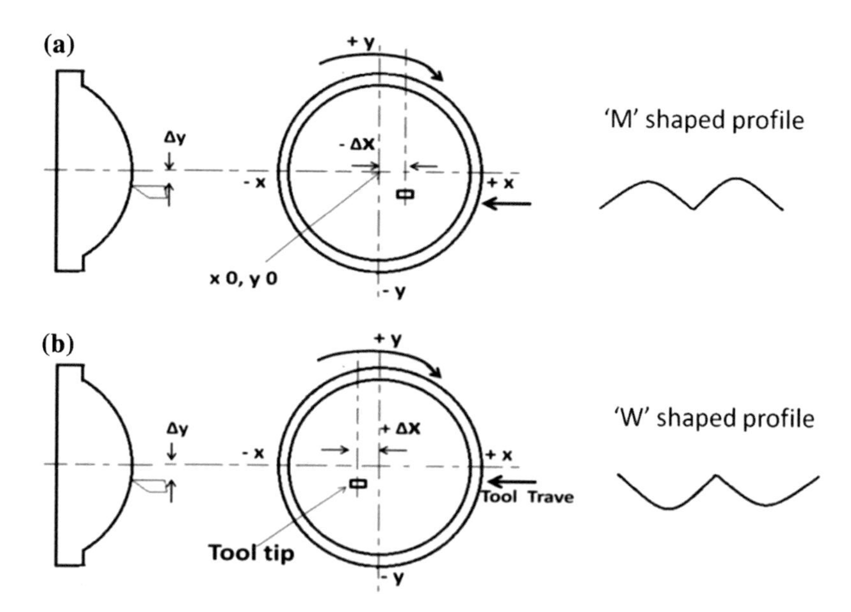

Fig. 10 Tool offset errors **a** positive X-offset, **b** negative X-offset

rotation axis is the x-offset and also known as ogive error [51]. The offset of tooltip in both x and y directions from rotational axis is shown in Fig. 10. For convex surface "M" and "W" shaped errors are formed if the tool is not reaching and crossing the center, respectively. The shapes of the error are opposite for the concave profile machining.

The vertical offset of the tool is called tool height error. For the zero rake angle tools, if tool is above the rotational axis, then a cone-shaped tip is formed at the center due to the impression of the front clearance of the tool flank face. However, if the tool is below the rotational axis of the workpiece, a cylindrically shaped tip is formed due to impression of zero rake face. For negative rake, in both the cases, cone-shaped tip is formed. Figure 11 shows height errors generated on convex spherical surfaces. The tool x-offset is the primary cause of the profile error. To minimize these errors, the tool tip should be aligned with axis of rotation. After height and x-offset corrections, tool nose radius correction is the next important task to achieve the desired profile. The tool nose radius should be corrected by equal and opposite magnitude of radius error of the fabricated surface.

6.2 *Component Mounting Fixtures*

The holding of workpiece is a crucial aspect of ultra-precision machining. It is important to hold the workpiece rigid enough to withstand against the cutting forces and at the same time to avoid any strain due to clamping. Mechanical clamping is

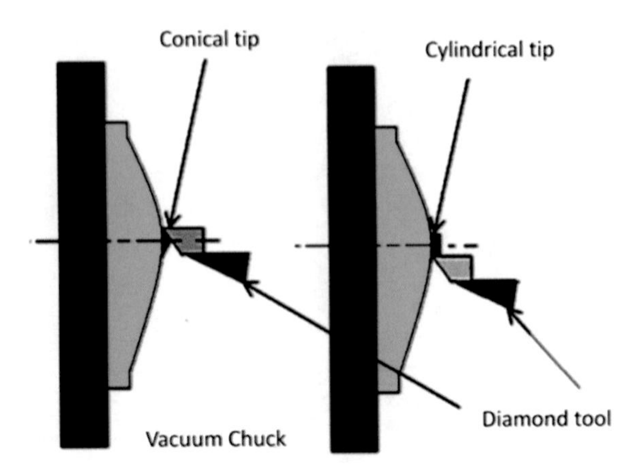

Fig. 11 Tool height error

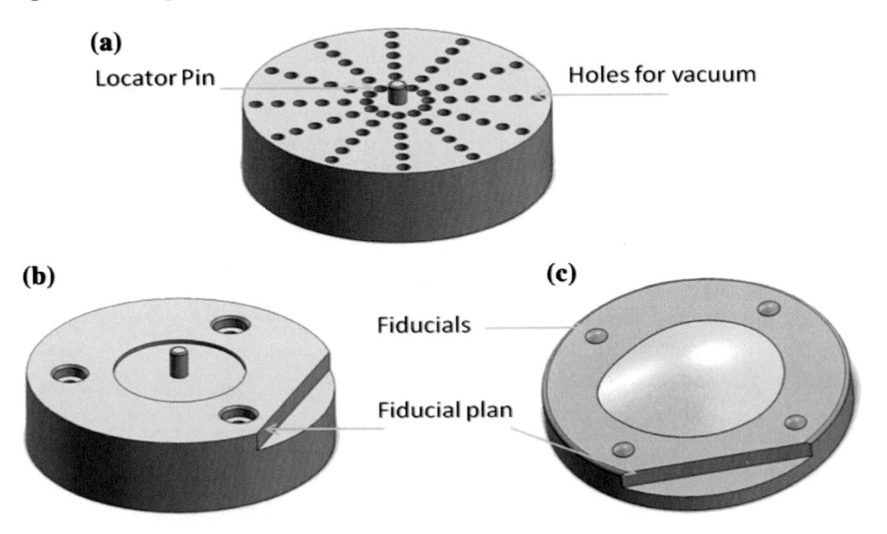

Fig. 12 Typical fixtures for ultra-precision machining

mostly avoided to protect the precise workpiece against any residual strain. Vacuum clamping is one of the most suitable methods to hold the components for ultra-precision machining [52]. Special purpose precise fixtures are required to hold the workpiece in vacuum chuck. First requirement of fixtures for vacuum chuck is to have a flat surface to rest on chuck surface to avoid any vacuum leakage. Vacuum then pass to the next surface of the fixture to hold the workpiece to be machined (Fig. 12a). For small and high aspect ratio components, vacuum clamping is also difficult. In such components, usually, special purpose wax is required to fix the component in fixture, which further can be held by vacuum clamping. Precise control of profile demands number of machining iterations. Effectiveness of compensation depends on remounting precision [53] It is easy to align and remount the rotationally symmetric components with good precision. However, for non-rotationally symmetric and micro-structured components, more degrees of freedom are required to restrict. Special purpose fixtures with number of fiducials are required for precise alignment and remounting of freeform surfaces. Precise alignment (submicron level) of freeform surfaces is still a challenging task [54, 55]. Typical fixtures for vacuum clamping of symmetric and freeform surfaces are shown in Fig. 12b, c.

6.3 Selection of Machining Parameters

High-quality surface finish can only be achieved by selection of optimal process parameters, i.e., tool feed rate, spindle speed, tool nose radius, proper coolant, and depth of cut [56, 57]. A typical case of machining parameters for aluminum 6061 machining is discussed to present the effects of process parameters. The trend is almost the same for most of the materials; however, optimum parameters can vary. The surface topography of the machined sample is considerably affected by the spindle speed and the effects are more dominant at the center. Coherence Correlation Interferometer (CCI 6000) measurement results of surface topography at the central part of the specimen (0.8 mm^2 area) are shown in Fig. 13. It can be seen that; the star patterns are prominent at lower spindle speeds i.e. 1000 and 2000 rpm. The star pattern is subsided at 3000 rpm, thus giving the best surface texture and roughness due to cancelation of tool-workpiece vibrations.

Tool feed rate leaves certain spiral marks on the workpiece during machining and these marks determine the high-frequency components of surface finish. The larger feed rate will lead to wider grooves or turning marks on the surface [58]. These marks deteriorate the surface finish and affect the performance of a system. Figure 14 shows the CCI image and SEM image of tool feed marks, respectively.

A residual mark of the tool on the surface depends on the tool feed rate and tool nose radius. Both are theoretically related by

$$R_t = f^2/8R$$

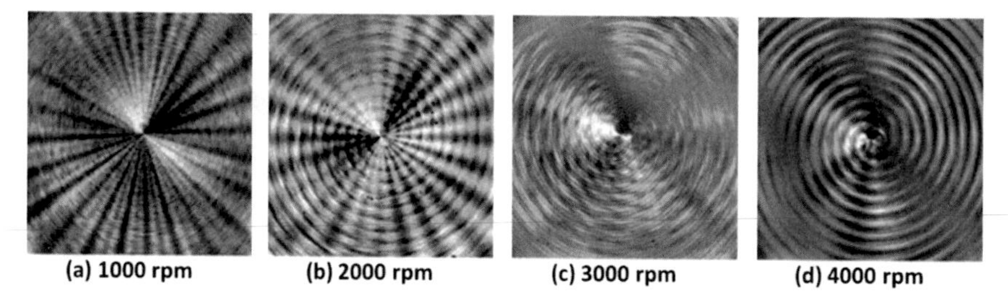

Fig. 13 Effect of spindle rotational speed on surface topology

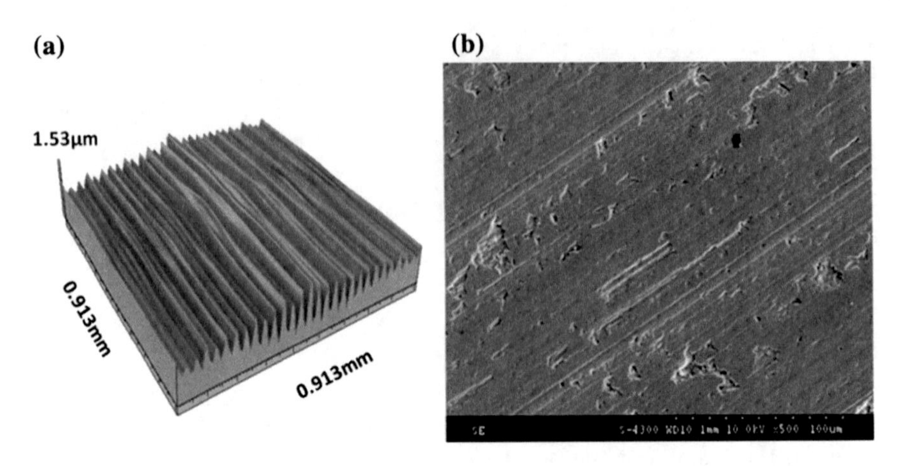

Fig. 14 **a** CCI image of diamond turned surface, **b** SEM micrograph showing tool feed marks

where R_t is highest peak-to-lowest valley difference in sample length, f is the feed rate, and R is tool nose radius. Figure 15 represents the surface roughness achieved at different spindle rotational speeds with two sets of tool nose radii: 0.5 and 1.0 mm for constant feed and depth of cut. The surface roughness is lower for higher tool nose radius. Spindle rotational speed of 3000 rpm delivers optimal surface roughness and surface profile.

Generally, depth of cut has minimal effect on surface roughness. However, the selection of suitable depth of cut is critical in case of brittle materials. Wrong selection of depth of cut may result in subsurface damage [59]. Low depth of cut is usually recommended to get good surface finish, to increase the tool life and to minimize the thermal issues. Optimum machining parameters for different materials are shown in Table 3.

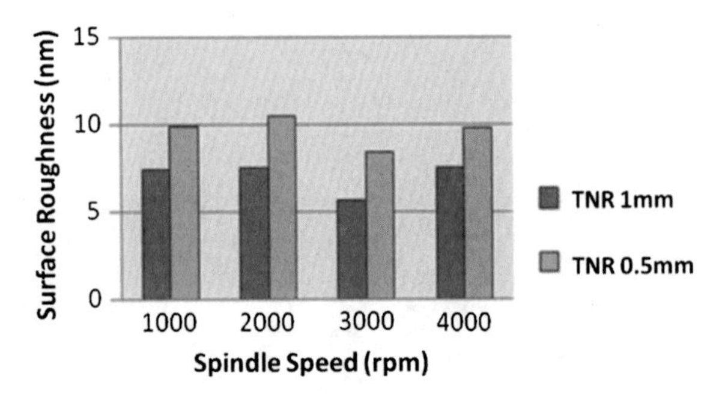

Fig. 15 Effect of tool nose radius on surface finish

Table 3 Suggested range of optimum machining parameters for different materials. (Reproduce with permission of Precitech) [60]

Material	Tool radius (mm)	Tool rake (in degree)	Front clearance (in degree)	Spindle speed	Depth of cut RPM (mm)	Feed rate finish pass (mm/rev)	Coolant finish pass
Germanium	0.635	−25	10	1000–2500	0.000508–0.00203	0.0013–0.002	Odorless mineral spirits (OMS) mist or stream
Zinc sulphide	0.635	−15	10	1000–2500	0.000508–0.00203	0.00254–0.00304	OMS, light oil mist or stream
Zinc selenide	0.635	−15	10	1000–2500	0.000508–0.00203	0.00254–0.00304	OMS, light oil mist or stream
Silicon	0.635	−25	10	1000–2000	0.000508–0.00203	0.00254	OMS, light oil mist or stream
Aluminum	0.508–1.524	0	10	1000–3000	0.00127–0.00254	0.00508	OMS, light oil mist
Copper	0.508–1.524	0	10	1000–3000	0.00127–0.00254	0.00508	OMS, light oil mist
Brass	0.508–1.524	0	10	1000–3000	0.00127–0.00254	0.00508	OMS, light oil mist
715 nickel alloy	0.508–1.524	0	10	1000–3000	0.00127–0.00254	0.00508	OMS, light oil mist
Electroless nickel	0.508–1.524	0	10	1000–3000	0.00127–0.00254	0.00508	OMS, light oil mist
PMMA plastic	0.508–1.524	0–5	15	2000–4000	0.0127–0.1016	0.0025–0.0063	Dry with air

6.4 Tool Path Generation and Compensation

The tool path is a representation of the required surface profile equation in the available machine coordinates which is generally done with the help of modern computer-aided manufacturing software. Tool path includes the information of machine coordinates, the designed surface and the dimensions of the cutting point of the tool, used to generate desired surface. The fabricated surface is characterized by a suitable metrology method, e.g., Interferometer or contact type profilometer for symmetric surfaces in general. The measurement data is compared with the design surface and used to correct the tool path to minimize the error footprint.

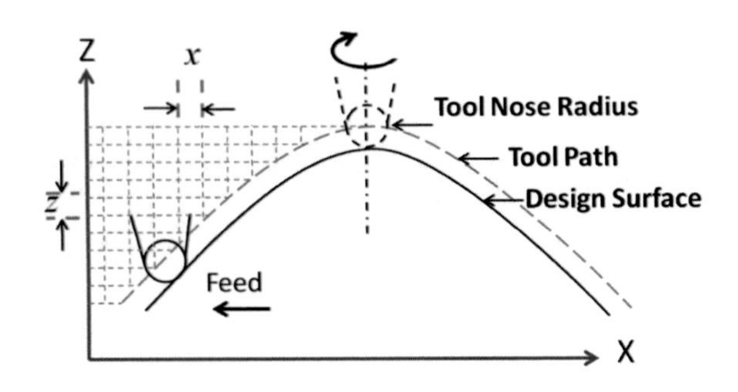

Fig. 16 Tool path generation for symmetric surface

6.4.1 Tool Path Generation for Symmetric Surface

Rotationally symmetric surfaces can be fabricated on two-axis (r, θ or x, z) configuration of ultra-precision machines. The spindle rotation is not programmed in this configuration and is set to optimum value. Synchronized movement of both x and x-axis is programmed by breaking the design equation into x and z coordinates with suitable increments [61]. The tool movement in z-axis will be the function of x-axis movement as per the programmed profile. For the spherical surfaces, apart from point to point movement, continuous movement can also be used with the help of common G and M programming codes. The compensation of tool nose radius is also incorporated into the tool path to avoid the form error due to offset of the programmed point and the actual cutting point [62], as shown in Fig. 16.

$$z = f(x)$$

6.4.2 Tool Path Generation for Non-rotationally Symmetric Surface

Fabrication of non-rotationally symmetric or freeform surfaces in ultra-precision machining is not possible in two-axis configuration. At least one additional axis is required to generate such profiles. Slow tool servo (STS) configuration of ultra-precision machining is a most suitable method to generate continuous type freeform surface [63]. In STS, the spindle rotation is synchronized with x and z-axis movements. The controlled rotation of the spindle is basically a C-axis, which restricts the independent spindle speed settings. To generate the tool path for freeform surfaces in STS machining, the design equation has to be transformed into cylindrical coordinates [64, 65]. A surface, which is the function of x and y coordinate is, can be expressed by sag value as

$$z = f(x, y)$$

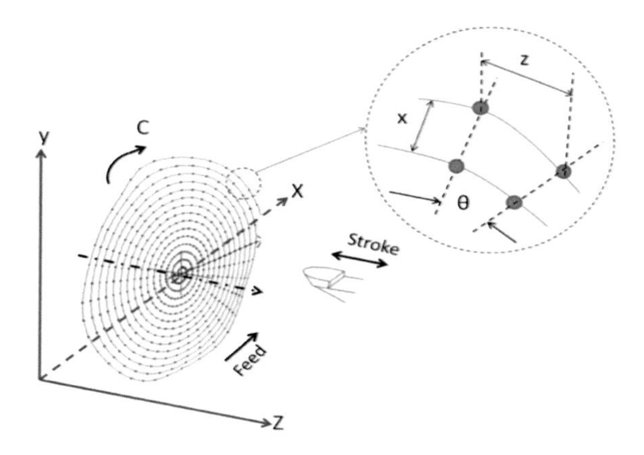

Fig. 17 Tool path generation for continuous freeform surfaces

for conversion into cylindrical coordinates

$$x = r\cos\theta$$

$$y = r\sin\theta$$

Therefore, in cylindrical coordinates

$$z = f(r, \theta)$$

where r is the radial position and θ is the angular position. The radial and angular position are defined on x-and c-axis, respectively as shown in Fig. 17. The tool movements for profile generation can be divided into suitable increments of radial (X-axis) and angular positions (C-axis). Tool nose radius compensation is more critical for freeform surfaces and makes the tool path generation more complicated [66].

6.4.3 Tool Path Generation for Micro-structured Surface

Due to the small size and non-rotationally symmetric distribution over the surface, the micro-structured surfaces are more difficult to fabricate than continuous type freeform surfaces. For generation of micro-structured surface, rapid strokes of the tool synchronize with relative C-axis position are required [67]. Due to heavy mass of slides, it is not possible in STS to programme the high-frequency strokes. Fast tool servo (FTS) is often used for this purpose [68]. FTS is the addition of secondary axis with Z-axis (known as W-axis) and enables the tool to reciprocate at high frequency with the help of piezo actuators. Three main steps are there to define the tool travel path for structured surfaces. First, to define the base surface over which micro-features are required. Second, to define the size and shape of micro-feature. Third, to define the distribution pattern of micro-features over the

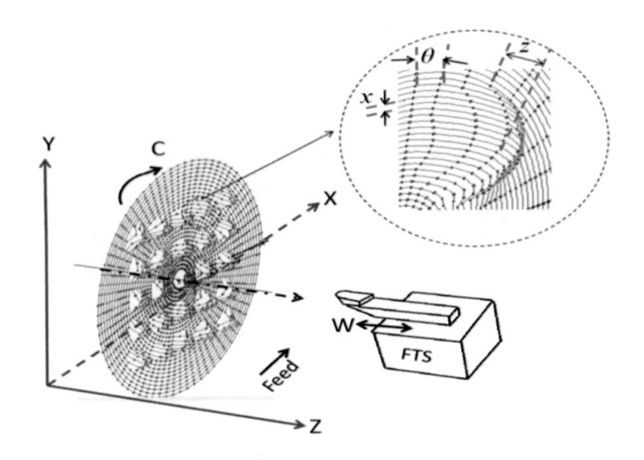

Fig. 18 Tool path generation for micro-structured surfaces

surface, which includes pitch, fill factor and pattern design, i.e., circular, rectangular, hexagonal, etc. [69]. The special purpose tools are often required to fulfil the requirements of size and shape of the micro-features. The movement of all the three axis (*C*, *X* and *Z*) is similar to the STS along with separately synchronized reciprocating tool along *Z*-axis [70], as shown in Fig. 18.

6.4.4 Tool Path Compensation

Surface after first fabrication cycle will always left with some errors which mainly include form and figure error. Magnitude of the error can be quantified by comparing the measured surface or features with the design surface. The measured data points are then analyzed for parameter of interest, i.e., form error, waviness [71, 72]. The error data points are added in the previous tool path to generate a modified tool path. This modified tool path is used to generate the required surface [73]. The compensation process is repeated till the desired surface quality is achieved. The profile error compensation cycle is shown in Fig. 19.

In case of freeform surfaces, compensation cycle breaks at the metrology point, due to unavailability of suitable metrology technique to give feedback to the fabrication process. Therefore, compensation of tool path to reduce the form error of freeform optics is even more complicated. Available metrology methods are not capable to directly measure the freeform surfaces and are still topic of interest for researchers [74–76]. Slope measurement technique by Shack–Hartmann sensor along with stitching method is also reported to measure the freeform surfaces [77–80]. However, the above metrology methods are still not explored to give the feedback to compensate for the tool path. Effective compensation is only possible by precise remounting of freeform optics with special purpose fixtures.

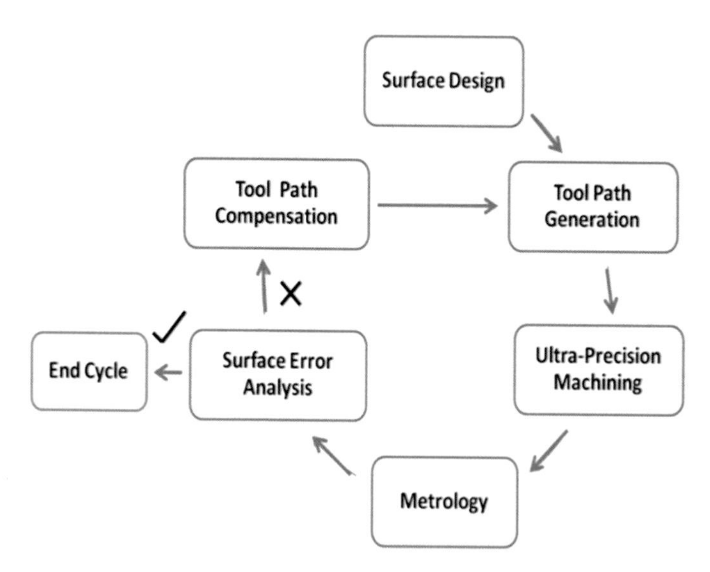

Fig. 19 Profile error correction cycle

6.5 *Effects of Uncontrollable Parameters*

Surface quality of the machined surface does not only depend upon input parameters but also on other factors, which are not in control of the operator, i.e., vibration during cutting, material spring back effect, dynamic balancing effects, and tool wear. Vibrations are always present in all machining processes and contribute to the surface roughness. The relative vibrations between the tool and the workpiece produce a sinusoidal pattern on the machined surface [81–83], as shown in Fig. 20a. Vibrations are produced due to many reasons, i.e., wrong selection of machining parameters, poor holding, tool wear, tool overhang, etc. [84]. Another main source of vibration is the dynamic unbalance of the machine spindle, which can cause the ripples or spindle star at the center (Fig. 20b). If the mass around the rotation axis is not same, resulting centrifugal forces can generate the significant vibrations at the center of the workpiece [85, 86].

The surface topography also depends on the structure of the material and its heat dissipation capabilities. Depending on its elastic properties and young's modulus, the material spring-backs to some amount and more commonly known as material swelling [87–89], as shown in Fig. 21a. This effect is more common in ductile materials like Al and Cu.

Diamond tool wear in ultra-precision machining is also a critical factor which not only degrades the surface but also increases the process cost significantly. The small amount of wear or microchipping of the cutting edge can increase the surface roughness drastically. Although, diamond is the hardest material, but it is difficult to avoid the tool wear completely. Wear mechanism of diamond tool is different for

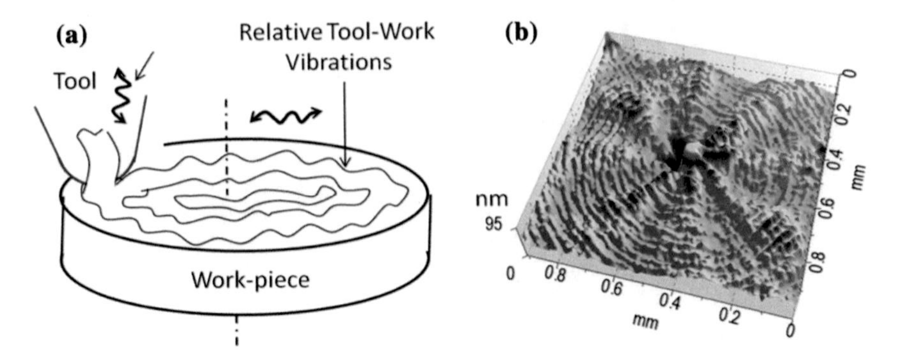

Fig. 20 **a** Relative tool-work vibrations, **b** effect of spindle unbalance

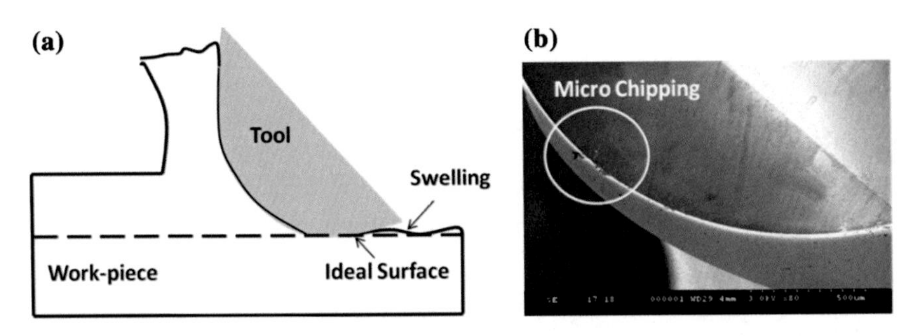

Fig. 21 **a** Material swelling effect, **b** microchipping of diamond tool edge

different materials [90]. Brittle materials offer more difficulty to process them, due to their hardness and results in rapid tool wear [47, 91–93]. The control of tool wear mainly depends on the machining parameter selection and understanding of tool wear mechanism. Wear measurement of diamond tool is essential to understand the wear mechanism, process behavior and tool life quantification. However, wear measurement of nano-metric cutting edge is one of the difficult tasks. SEM measurement is generally used to find the microwear or chipping of the cutting edge (Fig. 21b) [94, 95].

7 Metrology of Diamond Turned Surfaces

Metrology is a science of measurement and is an important part of any manufacturing process to control the quality of product and process at each stage. No manufacturing process is complete without qualification by a suitable metrology. The basic objectives of the metrology are to maintain the desired quality of product, to improve the process capability, to ensure the standardization and to reduce the overall cost of the process. Metrology is a wide area and covers all branches of engineering. However, current chapter is focused only on surface metrology.

Surface texture has an important role in functionality of different products for different applications. Surface texture error can be divided into two main categories, continuous and localized errors. The surface roughness, waviness and form are the global errors of surface texture and come under the category of continuous errors. All these errors are the result of unwanted effects of different variables of manufacturing process. The localized errors are the errors on small area as compare to the actual full surface [96]. For better understanding, surface texture errors of ultra-precision machined component can be divided into four orders (Fig. 22). First- and second-order errors or secondary texture mainly due to machine dynamics, which can be inaccuracies of machine, chatter due to vibrations, spindle in-balance, deformation of surface due to machining forces and thermal effects. These errors are of low to mid-level of frequency range and most commonly known as form error and waviness. The third- and fourth-order surface texture errors or primary texture is of high frequency and mainly due to tool-workpiece interaction [97, 98]. Tool feed marks, high-frequency vibrations due to cutting are generally termed as third-order surface irregularities, whereas surface damage during material separation is termed as fourth-order surface texture errors. Third- and fourth-order irregularities are most commonly measured as surface roughness.

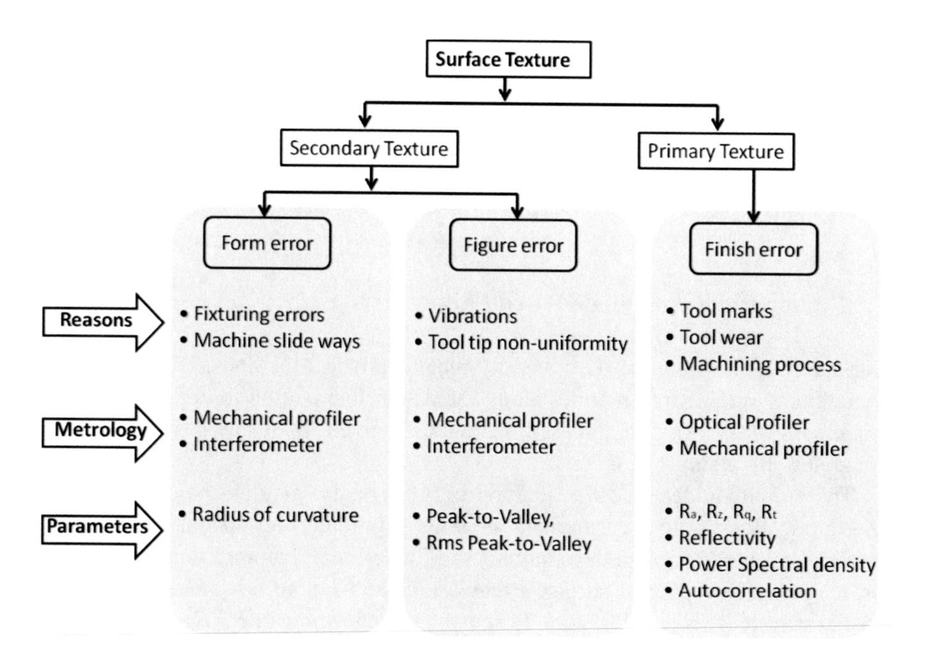

Fig. 22 Surface texture errors with their reasons

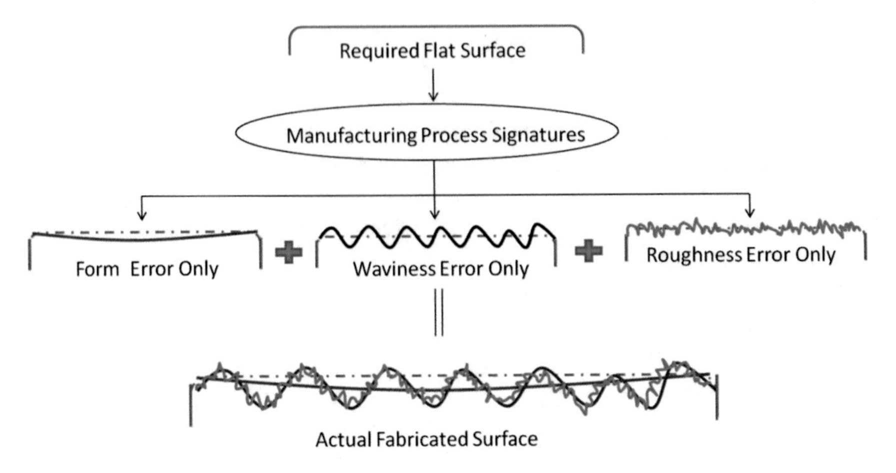

Fig. 23 Surface texture error types

Form error, waviness error, and surface roughness are always collectively present in the surface (Fig. 23). For quantitative measurement of one type of error, the other errors should be filtered for calculations.

7.1 Surface Finish

Surface roughness is a key parameter to define the surface finish, which decides the performance of product for many applications, i.e., tribological properties of mechanical and biomedical components, reflection and transmission for optical applications, fatigue life, and aesthetics of machine parts.

7.1.1 Important Amplitude-Based Parameters

Surface roughness can be described by many statistical methods. Historically, the variation of surface irregularities along the datum line is used to define the surface roughness [99]. Amplitude-based parameters given in Table 4 represents the roughness by a single number.

The amplitude-dependent evaluation of a roughness may not be enough for all the applications. The measurement of both amplitude and spatial interaction is required for more detailed description. Amplitude-based parameters are based only on heights and depths of surface irregularities and ignore any information about spatial wavelengths [100]. Figure 24 shows the same value of R_a, R_q, and R_t for two very different surfaces.

Table 4 Important amplitude-based parameters

Parameter	Definition	Mathematical representation
Arithmetic average roughness (R_a)	The arithmetic average of peaks and valleys on surface (R_a). It is the arithmetic average value of deviation of the profile from the mean reference line over the sampling length 'L'	$R_a = \frac{1}{L}\int_0^L \lvert z(x)\rvert\, dx$
Root-mean-square roughness (R_q)	R_q is root-mean-square of average roughness and it is more sensitive to random peaks and valleys	$R_q = \sqrt{\frac{1}{L}\int_0^L z^2(x)\, dx}$
Peak-to-valley roughness (R_t)	The total height from lowest valley to the highest peak is peak-to-valley (R_t) roughness and, it is useful in extracting the information of any sharp spike on the surface, such as scratch or crack	

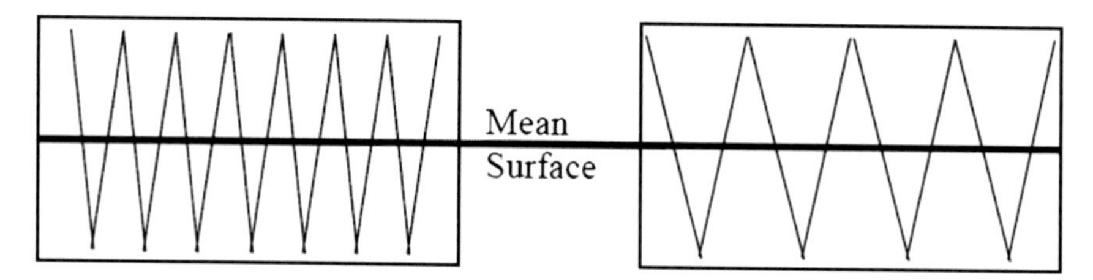

Fig. 24 Two different surface with same roughness

7.1.2 Frequency Domain Power Spectral Density Analysis

Randomly distributed peaks and valleys can be better represented by analyzing the spectral distribution, i.e., by looking the distribution of the roughness 'power' among the various surface frequencies (Fig. 25). To overcome the limitations of the amplitude-based parameters, the power spectral density (PSD) is used [101, 102].

The profile of the surface roughness of a machined surface includes tool feed marks, tool vibrations effect, material vibrations, and other material- and process-related effects [103]. The amplitude parameters, e.g., R_q, R_a or R_t are inadequate to provide this information about all these effects. Hence, for the better roughness analysis of high-quality surface, PSD has been preferred [104]. PSD gives the spatial frequency spectrum of the roughness measured in inverse-length units. Raw data can be extracted from the measurements of optical or mechanical profiler [105].

The power spectrum of a measured surface roughness profile is combination of various periodical components resulting from machining process. Mid- and low-order frequencies represent the waviness error and high-order frequencies represent the roughness of the surface. If the reason of particular frequency is known, process can be improved accordingly. The one-dimensional PSD can be

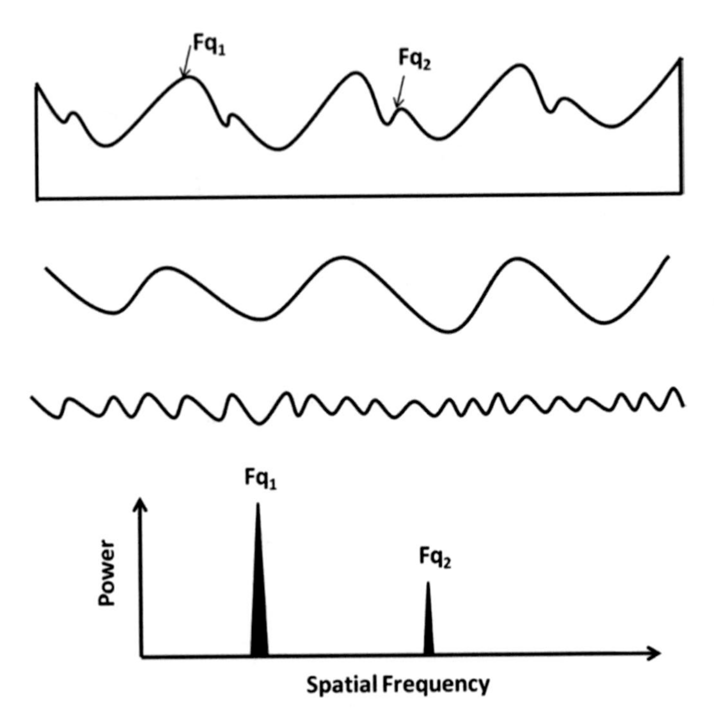

Fig. 25 Frequency distribution by PSD analysis

calculated from measured raw profile data cloud. If roughness height, $z(x$ is the function of distance 'x', a Fourier transform of finite-length may be written as

$$Z(k) = \frac{1}{L}\int_{0}^{L} z(x)\exp(-ikx)\,\mathrm{d}x$$

where k is the wave number and $i = \sqrt{-1}$. Practically, one can evaluate only a limited number of values of $z(x)$.

7.1.3 Autocorrelation

The autocorrelation is a mathematical tool utilized to study series or functions of values, i.e., time domain signals. The autocorrelation is effective to find recurring patterns in an available signal, i.e., finding the occurrence of a particular frequency which has been suppressed under unwanted frequencies [106]. For the random surfaces, the plot goes down to zero, whereas, for the surface has an influential spatial frequency, the plot fluctuates around zero in a periodic manner.

7.2 *Surface Form Characterization*

Aim of ultra-precision diamond turning is to achieve the nonmetric surface finish along with close agreement between the design and the archived form. Form of the machined surface is important to control for desired performance and it is more critical in some applications like optics. In simple words, form error is a deviation from desired or targeted surface profile. Form error of optical components is generally characterized with interferometric methods in terms of difference between ideal wavefront and distorted wavefront from fabricated surface [107, 108]. For fast measurements contact profilometer is also used to take two-dimensional scans at center [109, 110]. However, above methods are only capable to measure the form of rotationally symmetric surfaces. Form error measurement of freeform surfaces is still challenging and area of current interest. In case of structured surface size and position of micro-feature is an important parameter to measure. High-resolution microscopes, optical profilometer, scanning electron microscope, atomic force microscope, etc., are good enough to measure the shape, size, and position of micro-feature [111, 112].

7.3 *Surface Measurement Techniques*

The testing of ultra-precision machined surfaces mainly deals with the study of form error, figure error and surface finish (roughness) errors. The contribution of figure and form error is often collectively named as surface deviations. The root-mean-squares value (rms) and peak-to-valley value (Pv) are the frequently used single-value parameters to represent the surface roughness and surface deviations, respectively.

Many techniques are available to measure the surface quality of rotationally symmetric surfaces, such as mechanical profiling, white light interferometry, atomic force microscopy, and confocal laser scanning microscopy [113, 114]. Above techniques can be used to measure the roughness of freeform surfaces but failed to measure the form error. Metrology of freeform surfaces is still an area of research and many techniques are proposed in near past [20]. Also, research is going on to develop the Shack–Hartmann sensor based slope measurement technique to characterize the form error of freeform form surfaces [79, 80, 115]. Optical profilometer and high-resolution microscope, i.e., scanning electron microscope, atomic force microscope, etc., are most suitable for characterization of dimension and shape of micro-structures. Important surface measurement techniques are summarized in Table 5.

Table 5 Important surface measurement techniques and their relative advantages and limitations [116]

Metrology technique	Advantages	Limitation
Coordinated measuring machine (CMM)	• Can measure most of the shapes • Reflectivity of the surface/object is not a prerequisite	• Less accurate
Mechanical profiler	• Good sensitivity in Z-direction • Reflectivity of the surface/object is not a prerequisite	• Stylus scratch marks • Delivers only 2D profile
Interferometer	• Nondestructive in nature • Fast and accurate measurement	• Mainly for regular shapes • Expensive
Optical profiler	• Nondestructive in nature • Fast and accurate measurement	• Only for roughness • Expensive
Shack–Hartmann sensor	• Compact system • Less sensitive to vibrations	• Poor lateral resolution

8 Conclusion

The chapter covers the various steps of the ultra-precision machining process in detail. Important machine configurations and their applications for different areas are discussed. Overview of material removal mechanism for different category of engineering material is given. Various process parameters, i.e., tool setting; machining parameters, fixtures, tool path generation, and compensation are discussed. Importance of each step of process and related issues are highlighted. Current literature is aimed to create the understanding of ultra-precision machining process for generation of high-quality surface.

References

1. S.S. To, V.H. Wang, W.B. Lee, Single point diamond turning technology, in *Materials Characterisation and Mechanism of Micro-Cutting in Ultra-Precision Diamond Turning* (Springer, 2018), pp. 3–6
2. D. Huo, K. Cheng, Diamond turning and micro turning, in *Micro-Cutting* (Wiley Ltd, 2013), pp. 153–183
3. J. RAMSDEN, Description of an engine for dividing straight lines on mathematical instruments. Published by Order of the Commissioners of Longitude, J. Nourse, 1779
4. N. Ikawa, R. Donaldson, R. Komanduri, W. König, P. McKeown, T. Moriwaki, I. Stowers, Ultraprecision metal cutting—the past, the present and the future. CIRP Ann. Manuf. Technol. **40**, 587–594 (1991)
5. V. Mishra, G.S. Khan, K.D. Chattopadhyay, K. Nand, R.V. Sarepaka, Effects of tool overhang on selection of machining parameters and surface finish during diamond turning. Measurement **55**, 353–361 (2014)

6. C.F. Cheung, W.B. Lee, A theoretical and experimental investigation of surface roughness formation in ultra-precision diamond turning. Int J Mach Tool Manu **40**, 979–1002 (2000)
7. W.B. Lee, C.F. Cheung, J.G. Li, S. To, J.J. Du, Z.Q. Yin, Development of a virtual machining and inspection system for ultra-precision diamond turning. Proc. Inst. Mech. Eng. Part B J. Eng. Manuf. **221**, 1153–1174 (2007)
8. N. Khatri, R. Sharma, V. Mishra, M. Kumar, V. Karar, R.V. Sarepaka, An experimental investigation on the influence of machining parameters on surface finish in diamond turning of silicon optics, in *International Conference on Optics & Photonics 2015, International Society for Optics and Photonics* (2015), pp. 96540M-96540M-96548
9. Z. Li, C. Chan, W. Lee, Y. Fu, Spectral analysis of surface roughness and form profile of a machined surface after low pressure lapping. Proc. Inst. Mech. Eng. Part B J. Eng. Manuf. **230**, 1399–1405 (2016)
10. C. Wang, K. Cheng, N. Nelson, W. Sawangsri, R. Rakowski, Cutting force-based analysis and correlative observations on the tool wear in diamond turning of single-crystal silicon. Proc. Inst. Mech. Eng. Part B J. Eng. Manuf. **229**, 1867–1873 (2015)
11. F. Jafarian, D. Umbrello, S. Golpayegani, Z. Darake, Experimental investigation to optimize tool life and surface roughness in Inconel 718 machining. Mater. Manuf. Processes **31**, 1683–1691 (2016)
12. W. Sawangsri, K. Cheng, An innovative approach to cutting force modelling in diamond turning and its correlation analysis with tool wear. Proc. Inst. Mech. Eng. Part B J. Eng. Manuf. **230**, 405–415 (2016)
13. R. Sharma, N. Khatri, V. Mishra, H. Garg, V. Karar, Surface finish and subsurface damage distribution during diamond turning of silicon, in *Advanced Materials Proceedings*, ed. by H. Kobayashi (VBRI press, 2016), pp. 433–435
14. K.A. Desai, P.V.M. Rao, Machining of curved geometries with constant engagement tool paths. Proc. Inst. Mech. Eng. Part B J. Eng. Manuf. **230**, 53–65 (2016)
15. R.H. Abd El-Maksoud, M. Hillenbrand, S. Sinzinger, Parabasal theory for plane-symmetric systems including freeform surfaces. Opt. Eng. **53**, 031303 (2013)
16. X. Zhang, L. Zheng, X. He, L. Wang, F. Zhang, S. Yu, G. Shi, B. Zhang, Q. Liu, T. Wang, Design Fabric. Imag. Opt. Syst. Freeform Surf. **8486**, 848607–848610 (2012)
17. G.E. Davis, J.W. Roblee, A.R. Hedges, Comparison of freeform manufacturing techniques in the production of monolithic lens arrays, in *Proceedings SPIE* (2009), pp. 742605–742608
18. Y. Tohme, R. Murray, E. Allaire, Principles and applications of the slow slide servo, Moore Nanotechnology Systems White Paper (2005)
19. R.G. Ohl Iv, T.A. Dow, A. Sohn, K. Garrard, Highlights of the ASPE 2004 winter topical meeting on free-form optics: design, fabrication, metrology, assembly (2004), pp. 49–56
20. F.Z. Fang, X.D. Zhang, A. Weckenmann, G.X. Zhang, C. Evans, Manufacturing and measurement of freeform optics. CIRP Ann. Manuf. Technol. **62**, 823–846 (2013)
21. G. Chapman, Ultra-precision machining systems; an enabling technology for perfect surfaces, Moore Nanotechnology Systems (2004)
22. http://www.precitech.com/product/freeform/freeforml
23. http://www.nanotechsys.com/machines/
24. V.K. Jain, *Introduction to Micromachining* (Alpha Science International Limited, 2010)
25. R.L. Rhorer, C.J. Evans, Fabrication of optics by diamond turning, in *Handbook of Optics*, vol. 1 (1995), pp. 41.41–41.43
26. A.G. Thornton, J. Wilks, The wear of diamond tools turning mild steel. Wear **65**, 67–74 (1980)
27. R. Narulkar, S. Bukkapatnam, L.M. Raff, R. Komanduri, Graphitization as a precursor to wear of diamond in machining pure iron: a molecular dynamics investigation. Comput. Mater. Sci. **45**, 358–366 (2009)
28. F. Fang, X. Liu, L. Lee, Micro-machining of optical glasses—a review of diamond-cutting glasses. Sadhana **28**, 945–955 (2003)

29. E. Brinksmeier, W. Preuss, Micro-mach. Philos. Trans. Royal Soc. A Math. Phys. Eng. Sci. **370**, 3973–3992 (2012)
30. https://www.knightoptical.com/stock/optical-components/
31. G. Ghosh, A. Sidpara, P.P. Bandyopadhyay, Fabrication of optical components by ultraprecision finishing processes, in *Micro and Precision Manufacturing*, ed. by K. Gupta (Springer International Publishing, Cham, 2018), pp. 87–119
32. M.A. Davies, C.J. Evans, R.R. Vohra, B.C. Bergner, S.R. Patterson, Application of precision diamond machining to the manufacture of microphotonics components, in *Proceedings of SPIE* (2003), pp. 94–108
33. J. Wilks, Performance of diamonds as cutting tools for precision machining. Prec. Eng. **2**, 57–72 (1980)
34. http://www.ddk.com/PDFs/introtodiamondmachining.pdf
35. K. Ramesh, W.G. Lewis, S.C. Veldhuis, A. Yui, Redefining the diamond cutting edge: a technique that complements nano-metric surface generation. Mat. Manuf. Process. **20**, 895–903 (2005)
36. https://www.azom.com/article.aspx?ArticleID=12287
37. I. Durazo-Cardenas, P. Shore, X. Luo, T. Jacklin, S.A. Impey, A. Cox, 3D characterisation of tool wear whilst diamond turning silicon. Wear **262**, 340–349 (2007)
38. W.J. Zong, Z.Q. Li, T. Sun, K. Cheng, D. Li, S. Dong, The basic issues in design and fabrication of diamond-cutting tools for ultra-precision and nanometric machining. Int. J. Mach. Tools Manuf. **50**, 411–419 (2010)
39. D.A. Lucca, Y.W. Seo, R. Komanduri, Effect of tool edge geometry on energy dissipation in ultraprecision machining. CIRP Ann. **42**, 83–86 (1993)
40. R.K. Pal, H. Garg, R.V. Sarepaka, V. Karar, Experimental investigation of material removal and surface roughness during optical glass polishing. Mat. Manuf. Process. **31**, 1613–1620 (2016)
41. T.H.C. Childs, D. Dornfeld, D.E. Lee, S. Min, K. Sekiya, R. Tezuka, Y. Yamane, The influence of cutting edge sharpness on surface finish in facing with round nosed cutting tools. CIRP J. Manuf. Sci. Technol. **1**, 70–75 (2008)
42. M. Tauhiduzzaman, S.C. Veldhuis, Effect of material microstructure and tool geometry on surface generation in single point diamond turning. Prec. Eng. **38**, 481–491 (2014)
43. W.B. Lee, C.F. Cheung, S. To, A microplasticity analysis of micro-cutting force variation in ultra-precision diamond turning. J. Manuf. Sci. Eng. **124**, 170–177 (2002)
44. M.A. Rahman, M.R. Amrun, M. Rahman, A.S. Kumar, Variation of surface generation mechanisms in ultra-precision machining due to relative tool sharpness (RTS) and material properties. Int. J. Mach. Tools Manuf. **115**, 15–28 (2017)
45. M. Lai, X. Zhang, F. Fang, Crystal orientation effect on the subsurface deformation of monocrystalline germanium in nanometric cutting. Nanoscale Res. Lett. **12**, 296 (2017)
46. L. Chen, L. Hu, C. Xiao, Y. Qi, B. Yu, L. Qian, Effect of crystallographic orientation on mechanical removal of CaF2. Wear **376–377**, 409–416 (2017)
47. S.S. To, V.H. Wang, W.B. Lee, Machinability of single crystals in diamond turning, in *Materials Characterisation and Mechanism of Micro-Cutting in Ultra-Precision Diamond Turning* (Springer, Berlin, Heidelberg, 2018), pp. 43–69
48. G.P.H. Gubbels, Diamond turning of glassy polymers, PhD Dissertation. Eindhoven University of Technology, The Netherland (2006)
49. A. Baumgärtner, Statics and dynamics of the freely jointed polymer chain with Lennard-Jones interaction. J. Chem. Phys. **72**, 871–879 (1980)
50. V. Mishra, N. Khatri, K. Nand, K. Singh, R.V. Sarepaka, Experimental investigation on uncontrollable parameters for surface finish during diamond turning. Mater. Manuf. Process. **30**, 232–240 (2015)
51. M.C. Gerchman, Optical tolerancing for diamond turning ogive error, in *Reflective Optics II, International Society for Optics and Photonics* (1989), pp. 224–230
52. L.E. Chaloux, Part fixturing for diamond machining, in *28th Annual Technical Symposium, SPIE* (1984), p. 3

53. A. Sohn, Fixturing and alignment of free-form optics for diamond turning, in *Proceedings of the American Society for Precision Engineering Winter Topical Meeting on Free-Form Optics: Design, Fabrication, Metrology, Assembly, Citeseer* (2004)
54. M. Brunelle, J. Yuan, K. Medicus, J.D. Nelson, Importance of fiducials on freeform optics, in *SPIE Optifab, International Society for Optics and Photonics* (2015), pp. 963318-963318-963318
55. K. Medicus, J.D. Nelson, M. Brunelle, The need for fiducials on freeform optical surfaces, in *SPIE Optical Engineering + Applications, SPIE* (2015), p. 7
56. C.F. Cheung, W.B. Lee, Study of factors affecting the surface quality in ultra-precision diamond turning. Mater. Manuf. Process. **15**, 481–502 (2000)
57. V. Mishra, V. Karar, G.S. Khan, Analysis of surface roughness in slow tool servo machining of freeform optics. Asia Pac. J. (2017)
58. G.S. Khan, R.G.V. Sarepaka, K. Chattopadhyay, P. Jain, V. Narasimham, Effects of tool feed rate in single point diamond turning of aluminium-6061 alloy. Indian J. Eng. Mater. Sci. **10**, 123–130 (2003)
59. K.-W. Kim, A study on the critical depth of cut in ultra-precision machining. J. Korean Soc. Prec. Eng. **19**, 126–133 (2002)
60. http://www.precitech.com/technology-support/white-papers
61. H.-N. Cheng, Specifying optics to be made by single point diamond turning, https://wp.optics.arizona.edu/optomech/wp-content/uploads/sites/53/2016/12/Tutorial_ChengHN.pdf
62. X. Liu, L. Lee, X. Ding, F. Fang, Ultraprecision turning of aspherical profiles with deep sag, in *2002 IEEE International Conference on Industrial Technology, 2002. IEEE ICIT'02* (IEEE, 2002), pp. 1152–1157
63. Y.E. Tohme, J.A. Lowe, Machining of freeform optical surfaces by slow slide servo method, in *Proceedings of the American Society for Precision Engineering (ASPE) Annual Meeting* (2004)
64. C. Xu, K. Min, W. Xingsheng, H. Muhammad, Y. Jun, Tool path optimal design for slow tool servo turning of complex optical surface. Proc. Inst. Mech. Eng. Part B J. Eng. Manuf. **231**, 825–837 (2016)
65. V. Mishra, K. Pant, D.R. Burada, V. Karar, G. Khan, S. Jha, Generation of freeform surface by using slow tool servo, in *Freeform Optics, Optical Society of America* (2017), pp. FTh3B. 2
66. C.-C. Chen, C.-Y. Huang, W.-J. Peng, Y.-C. Cheng, Z.-R. Yu, W.-Y. Hsu, Freeform surface machining error compensation method for ultra-precision slow tool servo diamond turning, in *Proceedings SPIE* (2013), pp. 88380Y
67. M. Zhou, H.J. Zhang, S.J. Chen, Study on diamond cutting of nonrationally symmetric microstructured surfaces with fast tool servo. Mater. Manuf. Process. **25**, 488–494 (2010)
68. K. Rogers, J. Roblee, Freeform machining with precitech servo tool options, Precitech tutorials (2005)
69. D.P. Yu, S.W. Gan, Y. San Wong, G.S. Hong, M. Rahman, J. Yao, Optimized tool path generation for fast tool servo diamond turning of micro-structured surfaces. Int. J. Adv. Manuf. Technol. **63**, 1137–1152 (2012)
70. L. Qiang, Z. Xiaoqin, X. Pengzi, A new tool path for optical freeform surface fast tool servo diamond turning. Proc. Inst. Mech. Eng. Part B J. Eng. Manuf. **228**, 1721–1726 (2014)
71. W.B. Lee, C.F. Cheung, W.M. Chiu, T.P. Leung, An investigation of residual form error compensation in the ultra-precision machining of aspheric surfaces. J. Mater. Process. Technol. **99**, 129–134 (2000)
72. N. Khatri, V. Mishra, R.G.V. Sarepaka, Optimization of process parameters to achieve nano level surface quality on polycarbonate. Optimization **48** (2012)
73. S. Rohit, M. Vinod, K. Neha, G. Harry, K. Vinod, A hybrid fabrication approach and profile error compensation for silicon aspheric optics, in *Proceedings of the Institution of Mechanical Engineers, Part B: Journal of Engineering Manufacture* (2017) 0954405417733018
74. A.J. MacGovern, J.C. Wyant, Computer generated holograms for testing optical elements. Appl. Opt. **10**, 619–624 (1971)

75. X. Jiang, P. Scott, D. Whitehouse, Freeform surface characterisation-a fresh strategy. CIRP Ann. Manuf. Technol. **56**, 553–556 (2007)
76. J. Qiao, Z. Mulhollan, C. Dorrer, Optical differentiation wavefront sensing for freeform optics metrology, in *Frontiers in Optics, Optical Society of America* (2016), pp. FW5H. 5
77. K.K. Pant, D.R. Burada, M. Bichra, M.P. Singh, A. Ghosh, G.S. Khan, S. Sinzinger, C. Shakher, Subaperture stitching for measurement of freeform wavefront. Appl. Opt. **54**, 10022–10028 (2015)
78. G. Khan, Non-null technique for measurement of freeform wavefront using stitching approach, in *Freeform Optics, Optical Society of America* (2015), pp. FTh2B. 3
79. G. Khan, M. Bichra, A. Grewe, N. Sabitov, K. Mantel, I. Harder, A. Berger, N. Lindlein, S. Sinzinger, Metrology of freeform optics using diffractive null elements in Shack-Hartmann sensors, in *3rd EOS Conference on Manufacturing of Optical Components* (2013), pp. 13–15
80. D.R. Burada, K.K. Pant, V. Mishra, M. Bichra, G.S. Khan, S. Sinzinger, C. Shakher, Development of metrology for freeform optics in reflection mode, in *SPIE Optical Metrology, International Society for Optics and Photonics* (2017), pp. 103291K-103291K-103298
81. P.A. Meyer, A framework for enhancing the accuracy of ultra precision machining (2009)
82. S. Takasu, M. Masuda, T. Nishiguchi, A. Kobayashi, Influence of study vibration with small amplitude upon surface roughness in diamond machining. CIRP Ann. Manuf. Technol. **34**, 463–467 (1985)
83. C. Cheung, W. Lee, A theoretical and experimental investigation of surface roughness formation in ultra-precision diamond turning. Int. J. Mach. Tools Manuf. **40**, 979–1002 (2000)
84. A. Yip, Factors affecting surface topography in diamond turning (2014)
85. P. Huang, W.B. Lee, C.Y. Chan, Investigation of the effects of spindle unbalance induced error motion on machining accuracy in ultra-precision diamond turning. Int. J. Mach. Tools Manuf. **94**, 48–56 (2015)
86. Q. Wu, Y. Sun, W. Chen, G. Chen, Theoretical and experimental investigation of spindle axial drift and its effect on surface topography in ultra-precision diamond turning. Int. J. Mach. Tools Manuf. **116**, 107–113 (2017)
87. M. Kong, W. Lee, C. Cheung, S. To, A study of materials swelling and recovery in single-point diamond turning of ductile materials. J. Mater. Process. Technol. **180**, 210–215 (2006)
88. J. Kumar, V.S. Negi, K.D. Chattopadhyay, R.V. Sarepaka, R.K. Sinha, Thermal effects in single point diamond turning: analysis, modeling and experimental study. Measurement **102**, 96–105 (2017)
89. V. Mishra, A.K. Biswas, N. Kumar, L.M. Kukreja, R.V. Sarepaka, Fabrication of $\lambda/2$ phase step mirror for CO_2 laser resonator using diamond turning. Opt. Eng. **53**, 036107 (2014)
90. S. Zhang, S. To, G. Zhang, Diamond tool wear in ultra-precision machining. Int. J. Adv. Manuf. Technol. **88**, 613–641 (2017)
91. J. Yan, K. Syoji, J.I. Tamaki, Some observations on the wear of diamond tools in ultra-precision cutting of single-crystal silicon. Wear **255**, 1380–1387 (2003)
92. M.S. Uddin, K. Seah, X. Li, M. Rahman, K. Liu, Effect of crystallographic orientation on wear of diamond tools for nano-scale ductile cutting of silicon. Wear **257**, 751–759 (2004)
93. K. Singh, R.O. Vaishya, H. Singh, V. Mishra, S. Ramagopal, Investigation of tool life & surface roughness during single point diamond turning of silicon. Int. J. Sci. Res. **2**, 265–267 (2013)
94. M. Shi, B. Lane, C. Mooney, T. Dow, R. Scattergood, Diamond tool wear measurement by electron-beam-induced deposition. Prec. Eng. **34**, 718–721 (2010)
95. W. Gao, T. Motoki, S. Kiyono, Nanometer edge profile measurement of diamond cutting tools by atomic force microscope with optical alignment sensor. Prec. Eng. **30**, 396–405 (2006)

96. M. Maksimovic, Optical tolerancing of structured mid-spatial frequency errors on free-form surfaces using anisotropic radial basis functions, in *Optical Systems Design 2015: Optical Design and Engineering VI, International Society for Optics and Photonics* (2015), pp. 962613
97. http://what-when-how.com/metrology/meaning-of-surface-texture-and-some-definitions-metrology/
98. 1 The solid surface, in Tribology Series, ed. by I. Iliuc (Elsevier, 1980), pp. 1–20
99. D. Whitehouse, *Handbook of Surface and Nanometrology, University of Warwick* (Institute of Physics Publishing, Bristol and Philadephia, 2003)
100. G.S. Khan, R.G.V. Sarepaka, K. Chattopadhyay, P. Jain, R. Bajpai, Characterization of nanoscale roughness in single point diamond turned optical surfaces using power spectral density analysis. Indian J. Eng. Mater. Sci. **11**, 25–30 (2004)
101. G.S. Khan, *Characterization of Surface Roughness and Shape Deviations of Aspheric Surfaces*, PhD Dissertation, University of Erlangen-Nuremberg, Germany (2008)
102. J.K. Lawson, C.R. Wolfe, K.R. Manes, J.B. Trenholme, D.M. Aikens, R.E. English, Specification of optical components using the power spectral density function, in *Optical Manufacturing and Testing, International Society for Optics and Photonics* (1995), pp. 38–51
103. C.F. Cheung, W.B. Lee, A multi-spectrum analysis of surface roughness formation in ultra-precision machining. Prec. Eng. **24**, 77–87 (2000)
104. E. Marx, I.J. Malik, Y.E. Strausser, T. Bristow, N. Poduje, J.C. Stover, Power spectral densities: a multiple technique study of different Si wafer surfaces. J. Vacuum Sci. Technol. B: Microelectr. Nanometer Struct. Process. Measure. Phenom. **20**, 31–41 (2002)
105. J.M. Elson, J.M. Bennett, Calculation of the power spectral density from surface profile data. Appl. Opt. **34**, 201–208 (1995)
106. R.S. Sayles, T.R. Thomas, The spatial representation of surface roughness by means of the structure function: a practical alternative to correlation. Wear **42**, 263–276 (1977)
107. U. Griesmann, J. Soons, Q. Wang, D. DeBra, Measuring form and radius of spheres with interferometry. CIRP Ann. Manuf. Technol. **53**, 451–454 (2004)
108. A. Beutler, Metrology for the production process of aspheric lenses. Advanced Optical Technologies **5**, 211–228 (2016)
109. D.J. Whitehouse, Handbook of surface metrology (CRC Press, 1994)
110. R.V. Sarepaka, S. Sakthibalan, S. Doodala, R.S. Panwar, R. Kotaria, Surface characterization protocol for precision aspheric optics, in *Optifab 2017, International Society for Optics and Photonics* (2017), pp. 104481D
111. R. Scheuer, T. Mueller, E. Reithmeier, Development of a fast measurement system for microstructured surfaces, in *Imaging Systems and Applications, Optical Society of America* (2013), pp. JTu4A. 30
112. H.N. Hansen, K. Carneiro, H. Haitjema, L. De Chiffre, Dimensional micro and nano metrology. CIRP Ann. **55**, 721–743 (2006)
113. A. Duparre, J. Ferre-Borrull, S. Gliech, G. Notni, J. Steinert, J.M. Bennett, Surface characterization techniques for determining the root-mean-square roughness and power spectral densities of optical components. Appl. Opt. **41**, 154–171 (2002)
114. J.M. Bennett, Comparison of techniques for measuring the roughness of optical surfaces. Opt. Eng. **24**, 243380 (1985)
115. D.R. Burada, K.K. Pant, M. Bichra, G.S. Khan, S. Sinzinger, C. Shakher, Experimental investigations on characterization of freeform wavefront using Shack-Hartmann sensor. Opt. Eng. **56**, 084107 (2017)
116. A. Beaucampa, R. Freemana, R. Mortona, D. Walkerab, Metrology software support for free-form optics manufacturing, in *Proceedings Conference* (Chubu, Japan, Citeseer, 2007)

Part III
Micro and Nano Machining with Non Conventional Machining Techniques

Abrasive Waterjet Cutting of Lanthanum Phosphate—Yttria Composite: A Comparative Approach

K. Balamurugan, M. Uthayakumar, S. Sankar, U. S. Hareesh and K. G. K. Warrier

1 Introduction

Ceramic materials are well known for their thermal and physical properties such as high temperature withstanding capacity, high strength, and corrosion resistance which create an interest for the researchers. Due to the superior properties, rare earth phosphate materials particularly, Lanthanum Phosphate ($LaPO_4$) is found to be a suitable replacement material for a conventional material in aerospace applications. A monazite structure of $LaPO_4$ is identified to have very stable thermal properties and high melting points [1]. The wide application of rare earth phosphate material is limited, due to its poor toughness and high brittleness. Toughness is one of the major properties that have to be incorporated in ceramic materials to import machinability. $LaPO_4$ is preferred as an interface material. As the layered structure of $LaPO_4$ supports the formation of crack deflection and propagates crack while machining either by absorbing or by liberating fracture energy, it reduces crack tip [2]. Earlier studies reveal that stoichiometric $LaPO_4$ does not react with aluminum and zirconium oxides [3]. Addition of $LaPO_4$ in alumina composite imports an excellent machinable property [4]. A 30% of $LaPO_4$ in Al_2O_3-$LaPO_4$ composite reduces hardness and improves machinability [5]. $LaPO_4$-ZrO_2 composite gives

K. Balamurugan
Department of Mechanical Engineering, VFSTR (Deemed to be University), Vadlamudi, Guntur 522213, Andhra Pradesh, India

M. Uthayakumar (✉)
Faculty of Mechanical Engineering, Kalasalingam University, Krishnankoil 626 126, India
e-mail: uthaykumar@gmail.com

S. Sankar · U. S. Hareesh · K. G. K. Warrier
Material Sciences and Technology Division, National Institute for Interdisciplinary Science and Technology, Council of Scientific and Industrial Research, Thiruvananthapuram 695019, India

K. Kumar et al. (eds.), *Micro and Nano Machining of Engineering Materials*, Materials Forming, Machining and Tribology,
https://doi.org/10.1007/978-3-319-99900-5_5

high phase stability and is identified to be a non-reactive compound even at the high temperature of 1600 °C in the open atmosphere [6].

Among the studies made on different materials (CaO, MgO and Y_2O_3) upon developing a stabilizing Zirconia, Yttrium oxide (Y_2O_3) gives the least hardness value [7]. The transparent Y_2O_3 along with $LaPO_4$ finds its application in composite part because of its thermodynamic stability whose melting temperature is about 2410 °C [8]. The $LaPO_4$ and Y_2O_3 composites have been found to have good phase stability and most of the properties of Y_2O_3 and $LaPO_4$ are found to be identical in level [9]. They form a stabilized composite at room temperature and it makes them one of the suitable composites for high thermal barrier application [10].

In conventional methods, machining these composites is often difficult, due to their superior physical and mechanical properties. It requires a very high cutting energy and temperature. The removal of material in Abrasive Waterjet Machining (AWJM) is purely by erosion which is caused by the high-velocity abrasive particles. It is reported that no HAZ is generated on cut surface of two different materials welded through friction stir weld [11]. In AWJM technology, the kerf taper is characterized by a wider entry at the top than the exit at the bottom. It is one of the major barriers that limit its applications [12]. Early researchers had developed different methods and models to reduce the Kerf Angle (KA) in AWJM [13–15]. The effect of output performance characteristics on different operating conditions in AWJM with various abrasives on different materials shows diverse results [16–18]. Exclusive of input parameters, material properties and thickness of the materials predominantly determine the machinability of the material [19, 20]. Increase in stand-off distance and effect of jet pressure turn out on surface quality [21]. Despite, the review report on AWJ technology delivers that it is much suitable for polishing complex surfaces [22].

In the present work, the authors are very keen on knowing the effect of influence of using two different abrasives (Garnet and SiC) in AWJM with different operating conditions by varying the input parameters likely JP, SOD and TS on three output responses (MRR, KA, and Ra) on machining $LaPO_4 + Y_2O_3$ ceramic matrix composites. The microscopic examination study is performed on the machined surface to reveal mechanisms involved during the machining of this composite.

2 Materials and Method

2.1 Material Preparation

Defined stoichiometric mixture of lanthanum chloride and orthophosphoric acid are mixed with 25% ammonia as a supplementary product in an ultrasonic bath. A complete flocculated $LaPO_4$ is obtained at a pH of 6.8. The gained precipitate is rinsed several times in hot water followed by centrifuged force to remove the chlorine content. To convert the precipitate into a colloidal sol, 20% (Vol) of nitric

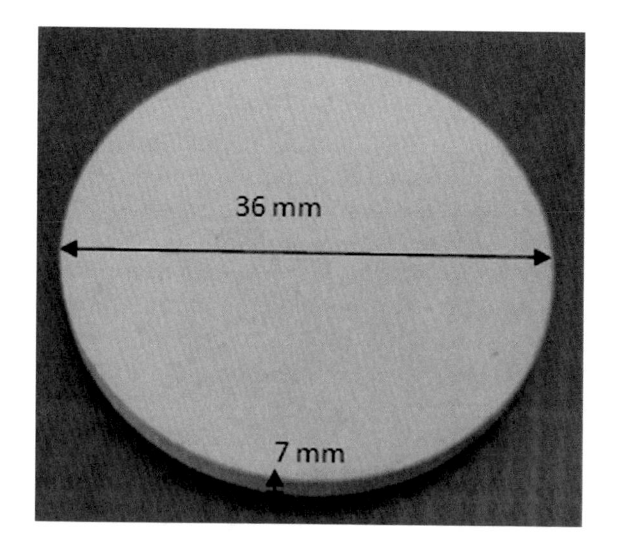

Fig. 1 Sample disk of $LaPO_4/Y_2O_3$ ceramic composite

acid is added with continuous stirring condition at pH in a range of 1.75–1.85. A calculated quantity of 99.9% pure yttrium nitrite hexahydrate is mixed with lanthanum phosphate sol. This composite precursor suspension is flocculated with the addition of ammonia with pH in the range of 7–8. The precipitate is stirred, dried and ball milled for 8 h. The gained powder is calcinated to 1400 °C for 2 h.

The obtained powder is consolidated to disk shape (36 mm diameter and 7 mm thickness) in titanium coated Oil Hardening Non-Shrinking Die Steel (OHNS) with a uniaxial compression force of about 480 MPa at room temperature. The sample disk is shown in Fig. 1.

During the compaction of $LaPO_4$ powder, the influence of high temperature has no significant effect on density [23]. Compressed samples are sintered in electric furnace for 2 h at 1400 °C. At elevated sintering temperature, the porosity of W-Y_2O_3 composite reduces with the increase of Y_2O_3 particles [24]. The properties of the prepared composite are shown in Table 1.

Table 1 Properties of $LaPO_4$-Y_2O_3 composite

S.no	Young's modulus (N/m^2)	Flexural strength (GPa)	Micro vickers hardness (GPa)	Theoretical density (g/cm^3)	Experimental density (g/cm^3)	Porosity (%)
1	4.96	96 ± 4	5.2	4.95	4.87	1.1616

2.2 Methods

AWJM of Model DIP 6D-2230 is used to cut the $LaPO_4/Y_2O_3$ ceramic composite prepared by Aqueous Sol-Gel process. An orifice of diameter 0.25 mm and WC nozzle of diameter 0.67 mm are used to cut the composite. Silicon Carbide and Garnet of 80 mesh size are used as abrasive particles. The Fig. 2 shows the experimental arrangements of AWJM. Table 2 shows the experimental condition of AWJM with three input factors and its levels.

The surface roughness tester of model SJ-411 that has a range of 350 μm with a probe speed of 0.25 mm/s over a span of 5 mm is used to measure the surface roughness on the kerf surface. The high precision weighing balance (AUX 220

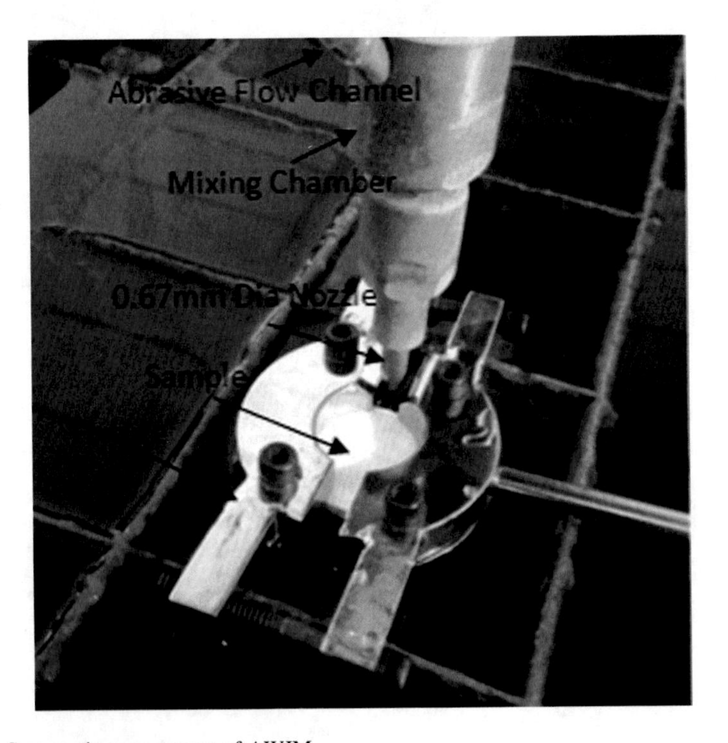

Fig. 2 Systematic arrangement of AWJM

Table 2 Selected factor and levels

S.no	Factors	Levels			Units
		1	2	3	
1	Jet pressure	220	240	260	bar
2	Stand-off-distance	1	2	3	mm
3	Traverse speed	20	30	40	mm/min

make of shimadzu) is used to measure the material loss before and after machining. It has a least count of 10 mg. The quantity of material removed per minute is calculated using the following Eq. (1).

$$MRR = \frac{(Wf - Wi) * 1000}{Dw * t} \quad (1)$$

where, W_i—Initial Weight of workpiece in grams before cutting, W_f—Final Weight of workpiece in grams after cutting, D_w—Density of the Workpiece (gm/cm^3) and t—Period of trial (s). Table 3 shows the experimental observations for Garnet and SiC abrasives for each output parameters.

Table 3 Experimental Observation

S.no	Input parameters			Garnet-output responses			SiC-output responses		
	JP (bar)	SOD (mm)	TS (mm/min)	MRR (g/s)	KA (Deg)	Ra (μm)	MRR (g/s)	KA (Deg)	Ra (μm)
1	220	1	20	0.02497	0.102	1.664	0.03189	0.215	1.191
2	220	1	30	0.02115	0.112	2.074	0.02589	0.315	1.328
3	220	1	40	0.01941	0.142	2.482	0.02152	0.364	1.486
4	220	2	20	0.02534	0.132	2.613	0.03258	0.258	1.139
5	220	2	30	0.02224	0.143	2.826	0.02998	0.322	1.289
6	220	2	40	0.01976	0.166	3.245	0.02681	0.371	1.542
7	220	3	20	0.02834	0.165	3.426	0.03471	0.298	1.344
8	220	3	30	0.02686	0.186	3.816	0.03186	0.342	1.508
9	220	3	40	0.02469	0.215	4.384	0.02895	0.398	1.675
10	240	1	20	0.02592	0.141	2.247	0.04112	0.292	1.231
11	240	1	30	0.02342	0.163	2.635	0.03412	0.364	1.494
12	240	1	40	0.02123	0.196	3.098	0.03097	0.411	1.651
13	240	2	20	0.02707	0.152	3.212	0.04526	0.293	1.331
14	240	2	30	0.02433	0.172	3.645	0.04291	0.362	1.531
15	240	2	40	0.02271	0.205	4.134	0.04017	0.434	1.682
16	240	3	20	0.03242	0.184	3.964	0.05621	0.425	1.515
17	240	3	30	0.02908	0.198	4.464	0.05014	0.468	1.656
18	240	3	40	0.02684	0.241	4.822	0.04519	0.517	1.797
19	260	1	20	0.02984	0.159	3.141	0.05495	0.348	1.423
20	260	1	30	0.02644	0.178	3.459	0.04951	0.396	1.572
21	260	1	40	0.02493	0.226	3.98	0.04652	0.492	1.692
22	260	2	20	0.03259	0.182	3.601	0.06488	0.393	1.598
23	260	2	30	0.03006	0.195	4.112	0.05715	0.462	1.731
24	260	2	40	0.02731	0.239	4.882	0.05359	0.554	1.881
25	260	3	20	0.03515	0.214	4.526	0.06887	0.516	1.645
26	260	3	30	0.03334	0.236	4.788	0.06574	0.598	1.847
27	260	3	40	0.03127	0.258	5.142	0.06252	0.681	1.966

2.3 Material Characterization

2.3.1 X-Ray Diffraction Study

The $LaPO_4/Y_2O_3$ powder extracted through Sol-Gel process is analyzed by powder X-Ray Diffraction method. Qualitative phase analysis is done by using the diffractometer D4 Endeavor Bruker AXS GmbHwith a θ-2θ geometry, operating at 50 kV and 30 mA with a Cu-Ka radiation (l = 1.5418 Å) in the range of 2θ = 10°–130°. The analyzation of the diffractogram is performed using the XRD-X'Pert high score software.

2.3.2 Micro-Structural Characterization Studies

To understand the mode of failure occurred in AWJM on $LaPO_4/Y_2O_3$ ceramic composite while using, two different abrasives of same mesh size are investigated through the kerf surface. The Energy Dispersive X-Ray Spectroscopy (EDS) study is performed to know the presence of elements in the composite.

3 Results and Discussion

3.1 X-Ray Diffraction

Phase transformation from rhabdophane to monoclinic type with hexagonal phase as an intermittent stage and the transformation of the elements to crystal is increased with the increase of sintering temperature [25]. The peak obtained in the composite is shown in Fig. 3. Monoclinic $LaPO_4$ and cubic Y_2O_3 are observed with an accompanying phase element of YPO_4. YPO_4 is found to be the dominating element with nearly 26% of contribution in the composite. It is believed that the high sintering temperature of 1400 ± 10 °C makesY_2O_3 and $LaPO_4$ molecules to get a new element of YPO_4. The solubility of Yttrium in Lanthanum Phosphate rises on increasing the sintering temperature [26]. Even at this elevated temperature of 1400 °C, $LaPO_4$ is found to be a non-reactive material with Y_2O_3. The surface morphology observed over SEM analysis reveals in the form of agglomerates and it is YPO_4. This compound helps the composite to protect from high-temperature oxidation and corrosion. Thus, this contribution of elements can be used for high-temperature applications.

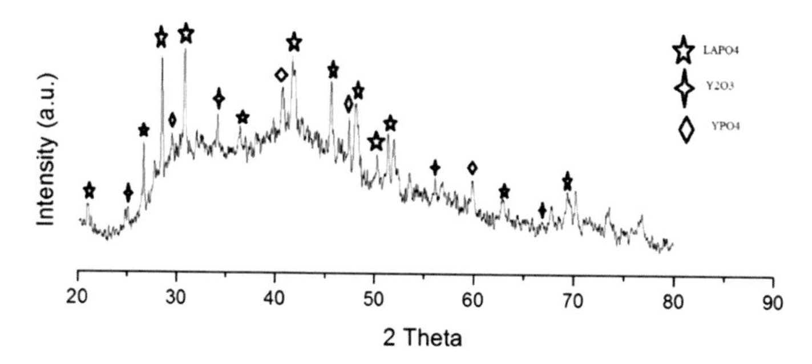

Fig. 3 XRD of $LaPO_4$ + 20% Y_2O_3 composite

3.2 *Energy Dispersive X-Ray Spectroscopy (EDS)*

The particle distribution with the spectrum of composite is shown in Fig. 4. The tabulated results provide a semi-quantitative view of the elemental composition in the fractured area in units of both weight percent and atomic percent. Well-dispersed elements of La, P, Y, and O are observed and the main composition of La is nearly 48 wt%, the contribution of P and Y of almost equal to 10 wt% and O of 30 wt%. From the XRD-X'Pert high score software analysis, it is identified that the significant addition of 20% Y_2O_3 in $LaPO_4$ composite are grouped to formulate compounds such as YPO_4 and Y_2O_3. High sintering temperature results in getting YPO_4 and Y_2O_3 with a composition wt% of 26 and 14.35, respectively.

3.3 *Effect of Abrasives on MRR*

MRR significantly interprets the manufacturing cost of the end product. Figure 5 shows the AWJM at varied operating conditions. For each successive pair, SOD and TS are set to constant and JP is the varying input parameter. It is compared with

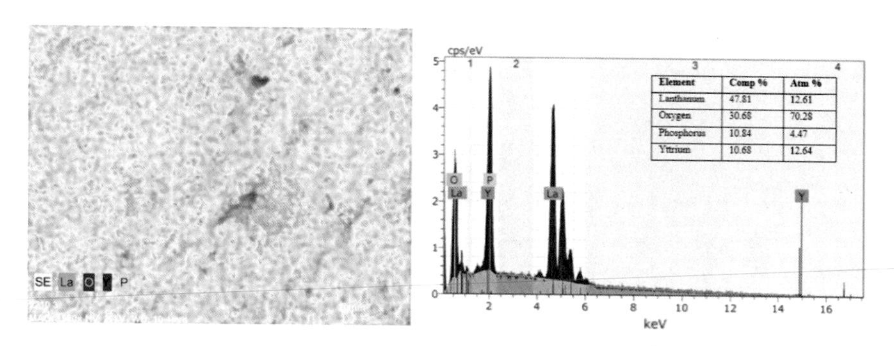

Element	Comp %	Atm %
Lanthanum	47.81	12.61
Oxygen	30.68	70.28
Phosphorus	10.84	4.47
Yttrium	10.68	12.64

Fig. 4 Elemental mapping and EDS spectra of 80 wt% $LaPO_4$ + 20 wt% Y_2O_3

MRR. Here, A1, A2, and A3 are three sets of operating conditions. With A1 at an operating condition of (JP = 220, 240 and 260 bar, SOD = 2 mm and TS = 20 mm/min), it is noticed that the rate of change of MRR for Garnet is considerably low with an increase in the JP. Whereas the similar types of observations are noticed in the operating conditions at A2 (JP = 220, 240, 260 bar, SOD = 2 mm and TS = 30 mm/min) and A3 (JP = 220, 240, 260 bar, SOD = 2 mm and TS = 40 mm/min). There is a significant increase in MRR on using SiC as abrasives and similar observations are recorded for the other two operating conditions. The high energy gained abrasives, when they impinge the hard surface, transfer the energy to the top layer element of the composite. SiC is found to have superior hardness over the Garnet and the rate of transfer of energy is high compared to Garnet. This increases the MRR in the composite. The wear rate and wear mechanism are highly influenced by the hardness of the abrasives [27].

From three operating conditions (A1, A2, and A3), it is recorded that there is a slight drop in MRR for both the abrasives. The increase in TS subsequently affects the bombardment rate of the abrasives and most likely reduces machining time consequences to lower the MRR.

In Fig. 6, the three operating conditions of B1 (JP = 220 bar, SOD = 1, 2 and 3 mm and TS = 30 mm/min), B2 (JP = 240 bar, SOD = 1, 2 and 3 mm and TS = 30 mm/min), and B3 (JP = 260 bar, SOD = 1, 2 and 3 mm and TS = 30 mm/min) are shown. JP and TS are set to constant by varying the SOD for each successive pair and they are compared with MRR.

The divergence at the first levels in three sets of observation has an increasing trend in MRR with the change in SOD. This is because of increase in JP. It is further noted that the effect of SiC abrasives is superior than garnet. The change in SOD on use of garnet on the three operating conditions (B1, B2, and B3) produces

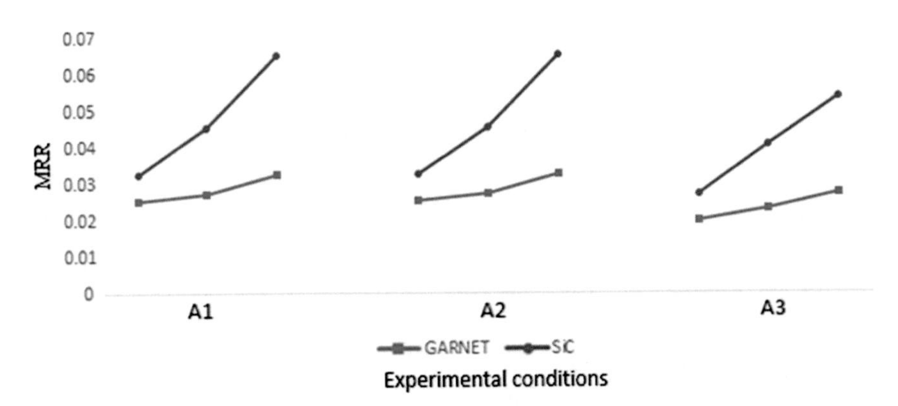

A1: JP=220, 240 and 260bar, SOD=2mm and TS=20mm/min
A2: JP=220, 240 and 260bar, SOD=2mm and TS=30mm/min
A3: JP=220, 240 and 260bar, SOD=2mm and TS=40mm/min

Fig. 5 Effect of JP and TS on MRR (g/s) with constant SOD

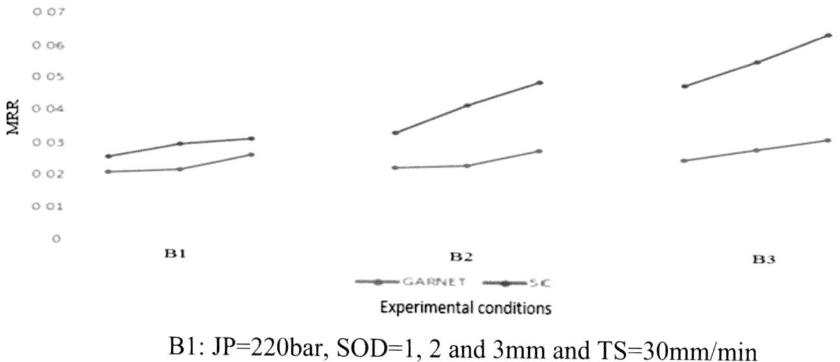

B1: JP=220bar, SOD=1, 2 and 3mm and TS=30mm/min
B2: JP=240bar, SOD=1, 2 and 3mm and TS=30mm/min
B3: JP=260bar, SOD=1, 2 and 3mm and TS=30mm/min

Fig. 6 Effect of JP and SOD on MRR with constant TS

a least significant because at constant JP and TS, the impinged high accelerated garnet irrespective of SOD losses some fraction of energy in breakage and forms new small grains. The hard SiC abrasive transfers most of its energy on machining the composite than that of garnet. This leads to having high MRR for SiC and it has a significant effect with increase in SOD on MRR. Increase in JP increases the impingement energy of the abrasives and which in turn increases backscatter abrasive particles energy. When it hits the cut surface consequences, it raises the MRR and it is noted on each pair of observation at different JP.

Figure 7 shows three pairs of observations: C1 (JP = 260 bar, SOD = 1 mm and TS = 20, 30, 40 mm/min), C2 (JP = 260 bar, SOD = 2 mm and TS = 20, 30 and 40 mm/min), and C3 (JP = 260 bar, SOD = 3 mm and TS = 20, 30 and 40 mm/min) of two different abrasives. Here, JP and SOD are set to constant with varied TS and they are compared with MRR for each successive pair.

Substantial reduction is observed in MRR on using of both the abrasives in all the three pairs with an increase in TS because the increase in TS reduces the number of bombardment and backscatter particles on the composite surface area considerably, due to lack of machining time. The MRR is found to be high in SiC because the random movements of hard SiC particles within the waterjet increase the width of jet beam and subsequently, it results in the removal of large amount of material in the composite.

3.4 *Effect of Abrasives on KA*

Figure 8 shows the three operating conditions in AWJM: A1 (JP = 220, 240, 260 bar, SOD = 2 mm and TS = 20 mm/min), A2 (JP = 220, 240, 260 bar, SOD = 2 mm and TS = 30 mm/min), and A3 (JP = 220, 240, 260 bar,

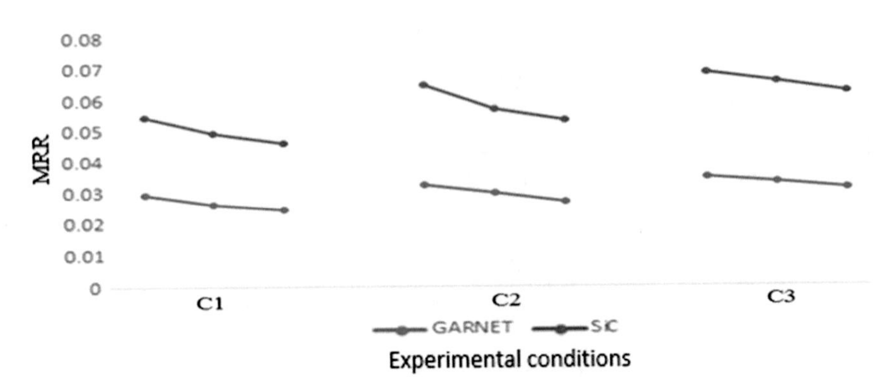

C1: JP=260 bar, SOD=1mm and TS=20, 30 and 40mm/min
C2: JP=260 bar, SOD=2mm and TS=20, 30 and 40mm/min
C3: JP=260 bar, SOD=3mm and TS=20,30 and 40mm/min

Fig. 7 Effect of SOD and TS on MRR with constant JP

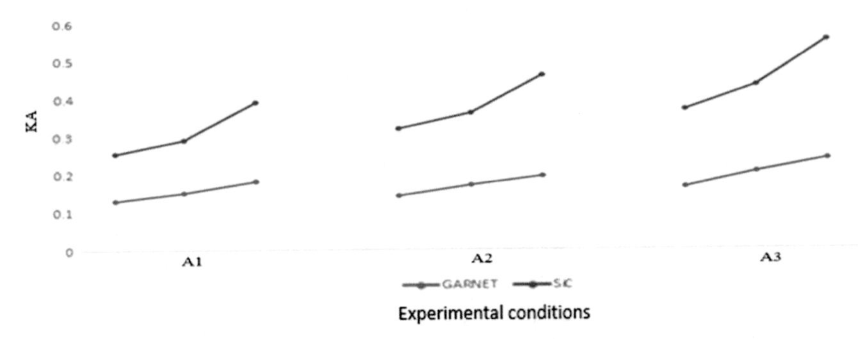

A1: JP=220, 240 and 260bar, SOD=2mm and TS=20mm/min
A2: JP=220, 240 and 260bar, SOD=2mm and TS=30mm/min
A3: JP=220, 240 and 260bar, SOD=2mm and TS=40mm/min

Fig. 8 Effect of JP and TS on KA with constant SOD

SOD = 2 mm and TS = 40 mm/min). Here, SOD and TS are set to constant with varied JP and they are compared with KA (Deg) for each successive pair. In all the observations, the kerf walls are converging from the top to the bottom with positive angles.

In all the observations, increasing JP results in a linear rate of change of KA in Garnet than SiC. The low hardness abrasives in all JP operating conditions behave in the same way by losing acceleration energy on machining the composite and by removing the excess material on the cut surface as backscatter particles. This gets increased with increase in JP and it results in the increase of cut surface region which provides larger top kerf width. The hard SiC particles, as they touch the top surface of the composite, impinge through and they flush to get a new top portion of the cut surface with the additional partial energy. The collision of backscatter particles with newly entered SiC abrasive results in getting the greater removal of material on the cut surface. This tends to increase KA. Considerable increase in KA for SiC occurs, due to the increase in JP. SiC tends to have high KA than Garnet because the impingement energy of SiC is found to be greater than Garnet, due to the less energy loss before it hits the sample and the hardness of the abrasive determines the rate of energy loss during the impact.

An increase in TS in each successive pair tends to have an increasing rate in KA. This happens because the increase in TS reduces the machining time which significantly reduces the abrasive flow rate over the kerf cut surface. This creates a wide entry at the top and narrow path at the bottom region of the composite.

In Fig. 9, three pairs of observations for two abrasives are shown. The three operating conditions are B1 (JP = 220 bar, SOD = 1, 2 and 3 mm and TS = 40 mm/min), B2 (JP = 240 bar, SOD = 1, 2 and 3 mm and TS = 40 mm/min) and B3 (JP = 260 bar, SOD = 1, 2, and 3 mm and TS = 40 mm/min). Here, JP and TS are set to constant with varied SOD and are compared with KA for each successive pair.

The change in SOD has a least significant effect on Garnet. The rate of change of KA on the use of Garnet is found to be less than SiC with respect to SOD. It is understood that the divergence of the jet with respect to SOD for Garnet is comparatively less than SiC. In SiC, low JP and change in SOD has least significant effect. At high JP, the bombarded abrasive particles, which possess superior hardness, as they are reflected after its first impact erodes a considerable amount of material on the kerf region that creates a slope and it results in an increase in KA.

Figure 10, shows the three pairs of observations: C1 (JP = 240 bar, SOD = 1 mm and TS = 20, 30, 40 mm/min), C2 (JP = 240 bar, SOD = 2 mm and

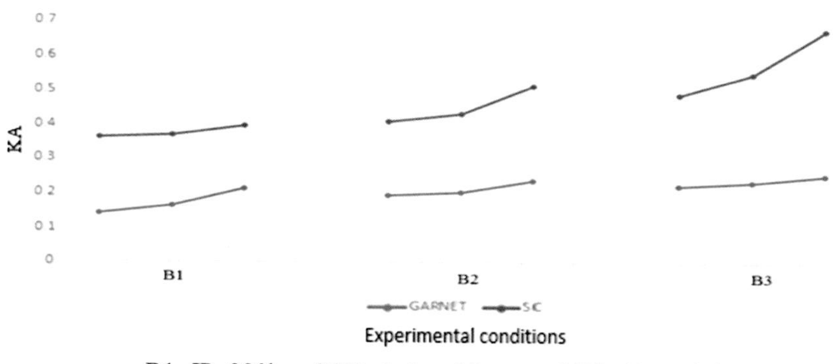

B1: JP=220bar, SOD=1, 2 and 3mm and TS=40mm/min
B2: JP=240bar, SOD=1, 2 and 3mm and TS=40mm/min
B3: JP=260bar, SOD=1, 2 and 3mm and TS=40mm/min

Fig. 9 Effect of JP and SOD on KA with constant TS

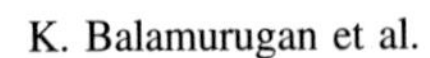

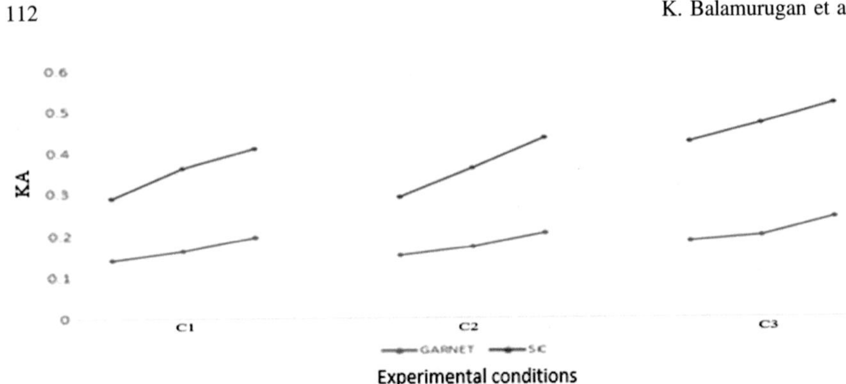

C1: JP=240 bar, SOD=1mm and TS=20, 30 and 40mm/min
C2: JP=240 bar, SOD=2mm and TS=20, 30 and 40mm/min
C3: JP=240 bar, SOD=3mm and TS=20,30 and 40mm/min

Fig. 10 Effect of SOD and TS on KA with constant JP

TS = 20, 30 and 40 mm/min) and C3 (JP = 240 bar, SOD = 3 mm and TS = 20, 30 and 40 mm/min) of two different abrasives. Here, JP and SOD are set to constant with varied TS and are compared with MRR for each successive pair.

KA observations tend to increase with increase in TS. The rate of change in KA for Garnet and SiC abrasive has similar effect on the first two pair of observations, For the third, a slightly increased observation is recorded because with increase in SOD, the width of the jet beam increases and it causes a partial loss of energy gained by the abrasives before hits the surface. As the high acceleration SiC particles, when they hit the composite, they lose their energy on machining and it is believed that the energy possessed by the SiC as a backscatter particle has high bouncing effect after it hits the target.

3.5 *Effect of Abrasive Particles on Ra*

Figure 11 shows the Ra observation under three cutting conditions in AWJM. They are A1 (JP = 220, 240, 260 bar, SOD = 3 mm and TS = 20 mm/min), A2 (JP = 220, 240, 260 bar, SOD = 3 mm and TS = 30 mm/min) and A3 (JP = 220, 240, 260 bar, SOD = 3 mm and TS = 40 mm/min). SOD and TS are set to constant with varied JP and are compared with Ra.

Increase in JP significantly affects the surface finish of the composite. Increase in JP consequently increases the abrasive flow rate in the waterjet which results in an increase of the width of the waterjet, due to the unsystematic movement of the particle. This effect consequently produces a superior surface finish irrespective of MRR and KA. It is noticed that Ra of SiC is lesser than the Garnet because the utilization of the acceleration energy gained through the high-pressure water for

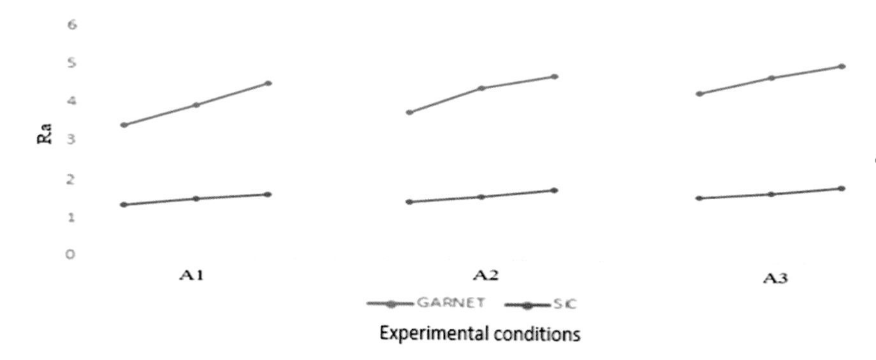

A1: JP=220, 240,260bar, SOD=3mm and TS=20mm/min
A2: JP=220, 240,260bar, SOD=3mm and TS=30mm/min
A3: JP=220, 240,260bar, SOD=3mm and TS=40mm/min

Fig. 11 Effect of JP and TS on Ra with constant SOD

machining the composite is high in SiC. The increasing pressure increases the average width of cut on the use of hard SiC as abrasives [28].

Figure 12 shows the Ra observations in three operating conditions: B1 (JP = 220 bar, SOD = 1, 2 and 3 mm and TS = 30 mm/min), B2 (JP = 240 bar, SOD = 1, 2 and 3 mm and TS = 30 mm/min) and B3 (JP = 260 bar, SOD = 1, 2 and 3 mm and TS = 30 mm/min). Here, JP and TS are set to constant with varied SOD and are compared with Ra for each successive pair.

The rate of change in Ra for SiC is found with uniform increase which lies in the acceptable range. It shows that SOD has least significant in the determination of Ra. Similar observations are noted for Garnet. The rate of increase of Ra is high in SiC

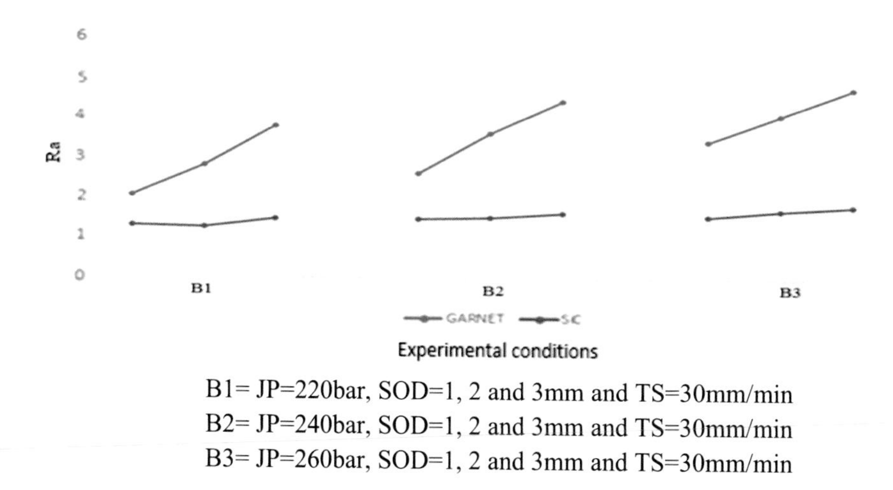

Fig. 12 Effect of JP and SOD on Ra with constant TS

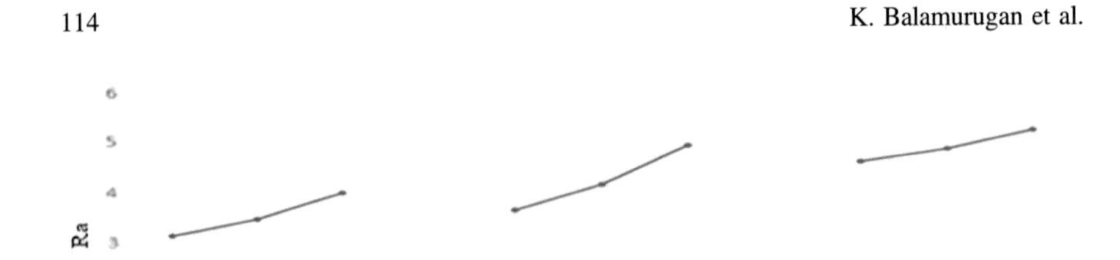

C1: JP=260 bar, SOD=1mm and TS=20, 30 and 40mm/min
C2: JP=260 bar, SOD=2mm and TS=20, 30 and 40mm/min
C3: JP=260 bar, SOD=3mm and TS=20, 30 and 40mm/min

Fig. 13 Effect of SOD and TS on Ra with constant JP

and it reveals the significance of JP irrespective of SOD. The acceleration energy of the SiC reduces significantly, due to the collision of particles within itself and creates a wider water beam which results in the removal of excess amount of material leaving wear scar and track on the kerf surface.

Figure 13 shows the three pairs of observations: C1 (JP = 260 bar, SOD = 1 mm and TS = 20, 30, 40 mm/min), C2 (JP = 260 bar, SOD = 2 mm and TS = 20, 30 and 40 mm/min) and C3 (JP = 260 bar, SOD = 3 mm and TS = 20, 30 and 40 mm/min). Here, JP and SOD are set to constant with varied TS and are compared with Ra for each successive pair.

Increase in TS considerably determines the Ra for both the abrasives. High TS leads to having a large kerf angle difference and results in the formation of striation. It considerably reduces the surface waviness on the kerf surface. Surface roughness is found to increase with high TS and this significantly reduces the surface flaws [29].

3.6 *Influence of Process Parameters Over Different Abrasives*

The MRR and KA are found to be superior to Garnet at all working conditions. When compared to Garnet, SiC produces superior surface finish. In all working conditions, the impingement force gets increased with increase in JP, the width of the water beam increases with increase in SOD and increase in TS subsequently reduces the machining time which significantly affects the output responses. The hard SiC abrasive particle removes a large amount of materials in the composite. Because of high JP, the materials are removed in large number by the processes of high impingement force, wider water beam, and high energy backscatter abrasives. In spite of excess removal of materials SiC produces a better surface finish. Increase in SOD increases the width of the beam that creates a larger kerf top surface and smaller kerf bottom surface. It results in an increase of KA. Comparatively, SiC produces larger KA than Garnet in all working environments. Garnet after the impingement loses its energy by transferring, forces to a composite material for machining. Additionally, energy drops take place due to the fracture of abrasive on impact load. These small, partial energy gained particles remove a negligible portion of materials in the kerf surface by creating a lesser kerf taper angle. The backscatter particles create micro scar and produce rough cut at the bottom surface of the composite.

4 Micro-Structural Characterization Study

4.1 Characterization Studies on SiC Machined Surfaces

Surface integrity determines the quality of the end product. The AWJ machined $LaPO_4/Y_2O_3$ composite on micro-structural characterization analysis on the kerf wall is shown in Fig. 14. Figure 14a shows a bulk material removal through a weak grain boundary. Crack extension.

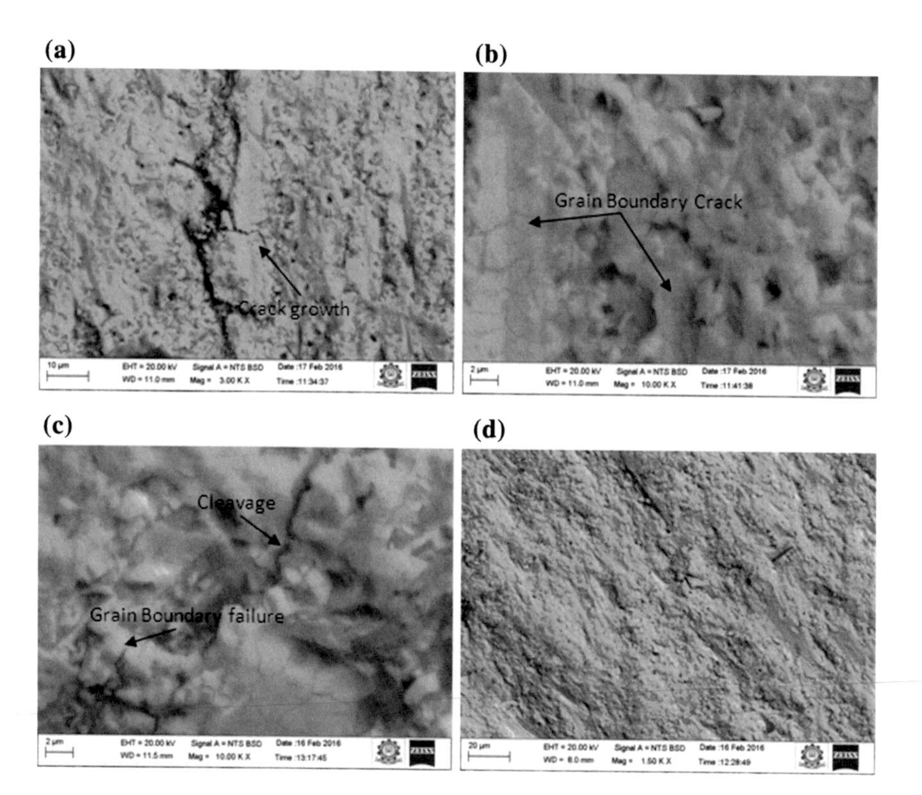

Fig. 14 a–d Observations on micro structure characterizations along kerf wall on use of SiC as abrasives

The impinge force of SiC apart from removal of composite tends to rise the temperature on the cut surface and gets cooled by water on the removal of material. This action induces thermal stress. It may weaken the grain boundary and cause the propagation of microcrack. Subsequent rise and drop in temperature on specimen surface in AWJM is conducive to residual stresses and is expected to contribute micro-cracking [30]. The thermal stress induced over the grain boundary is noticed on cut surface and is shown in Fig. 14b. The crack initiation, propagation, and the particle about to deform from the plane are shown in Fig. 14c. Both the trans-granular and intergranular failures are observed over the surface. Figure 14d shows the direction of abrasive jet flow. In general, it is believed that the hardness of SiC abrasive regulates grain fracture along the grain boundary and also through the grains. Plastic deformation surface is obtained over the cut surface of the composite.

4.2 Characterization Studies on Garnet Machined Surfaces

Figure 15 shows the micro-structural characterization of cut surface. In Fig. 15a, the micro level cracks that are formed along the grain boundary are visualized. Few microcracks are seen over the kerf surface and they are the evidence that the heat generated during machining is less in Garnet than SiC. This low-temperature rise and subsequent cooling produce less thermal distortion than SiC. The high accelerated Garnet abrasives import its energy to the composite on machining which results in squeezing and crushing of composite material. They lead to gain a new kerf surface.

Figure 15b shows the rough cut region at high JP. The collision of Garnet within itself in the water beam before it hits the target partially, losses its shape and create jet divergence. These low hardness abrasives, while on machining the composite shutter into small pieces, result in ploughing effect and mark wear track on the cut surface at large micron level. The abrasive wear mechanism mainly depends on the hardness of the cutting material and the cut material. While using Garnet in high working environment, the fully energy gained abrasives impact over this region and it is reduced considerably due to TS. This results in the formation of striations, large crater and overlapping region over the kerf surface. It is shown in Fig. 15c. Figure 15d shows the top kerf surface of the composite. Literally, an acceptable level of surface finish is obtained over this surface. When abrasive passes through the composite material because of its low hardness, it breaks to get fine grain. These grains act as backscatter abrasives and they flush out the top kerf surface material by increasing KA.

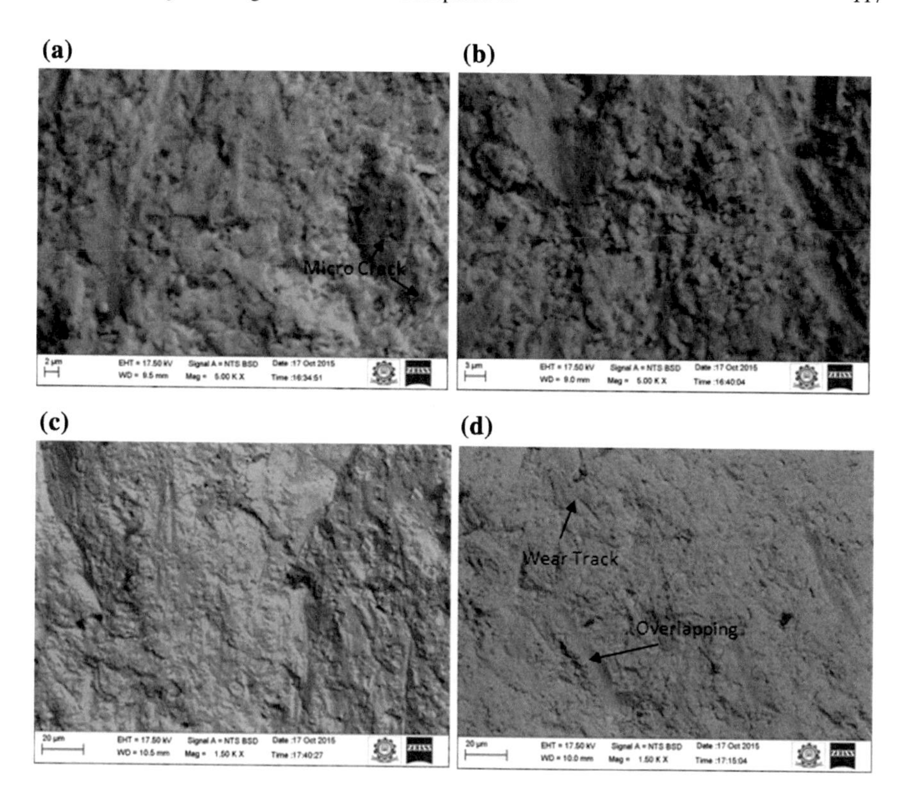

Fig. 15 a–d Micro structure characterizations along kerf wall on use of Garnet as abrasives

5 Conclusion

The effects of the influence of input parameters over each output response on different operating conditions using Garnet and SiC as abrasives in AWJM on $LaPO_4/Y_2O_3$ ceramic matrix composite are investigated and it reveals the following:

- MRR is found to be high for SiC. The mechanism behind this is that SiC abrasive gains high kinetic energy through JP irrespective of SOD and TS impinges the composite surface with high energy. The hard SiC transfers its energy to machine the composite and bounces back with great energy. With the increase of depth of cut, the bounced abrasives start hitting the kerf surface and create a wider kerf taper. This does not happen in Garnet because at high force when the Garnet hits the composite, it breaks into pieces. Due to the low hardness, considerable amount of the accelerated energy gained through JP is wasted besides machining. These grains when they act as backscatter abrasives create wear track and ploughing effect on the kerf surface of the composite.

- KA is found to be greater for SiC because the high energy possessed by the backscatter abrasives after hitting the primary region bounce back and with force greater than Garnet particles, they slide through the cut surface with increase in JP. The increase in SOD considerably increases the width of the water beam which causes a shift in observation and high TS significantly increases the KA, due to the quick movement of the nozzle over the composite. This reduces the abrasive flow rate over the bottom cut region.
- Ra is found to be superior in SiC than Garnet. Hardness of abrasives determines the surface quality of this composite. These hard abrasives gain kinetic energy from the high-speed water and this energy is utilized completely for machining. Apart from machining, some part of the energy is wasted by shuttering and new fine grains are formed, while Garnet is used as abrasive.
- Microscopy examinations of the kerf surface of the composite reveal that on SiC machined surface, the failure occurs by grain boundary deformation. The developed internal stress and repetitive cyclic impact load of hard SiC abrasives over the composite lead to get plastic deformation. In Garnet, a few microcracks form, due to the crushing and squeezing of composite particle and machining is done by erosion of materials. The formation of microcrack on the cut surface happens by induced thermal stress. More numbers of microcracks are formed in SiC and the machined surface reveals that the working temperature of SiC is significantly higher than Garnet.

Acknowledgements The authors wish to express their thanks to DST-FIST Sponsored Advance Machining and Measurement Laboratory-Kalasalingam University for their support rendered to pursue this research work.

References

1. S. Lucas, E. Champion, D.B. Assollant et al., Rare earth phosphate powders $RePO_4 \cdot nH_2O$ (Re = La, Ce or Y) II Thermal behavior. J. Solid State Chem. **177**, 1312–1320 (2004)
2. G. Gong, B. Zhang, H. Zhang et al., Pressure less sintering of machinable $Al_2O_3/LaPO_4$ composites in N_2 atmosphere. Ceram. Int. **32**, 349–352 (2006)
3. P.E.D. Morgan, D.B. Marshall, R.M. Housley et al., High-temperature stability of monazite-alumina composites. Mater. Sci. Eng. A **195**, 215–222 (1995)
4. R. Wang, W. Pan, M.J. Chen et al., Properties and microstructure of machinable $Al_2O_3/LaPO_4$ ceramic composites. Ceram. Int. **29**, 19–25 (2003)
5. M. Abdul Majeed, L. Vijayaraghavan, S.K. Malhotra et al., Ultrasonic machining of Al_2O_3/$LaPO_4$ composites. J. Mach. Tools Manuf. **48**, 40–46 (2008)
6. W. Min, K. Daimon, T. Matsubara et al., Thermal and mechanical properties of sintered machinable $LaPO_4$-Zr_2O_2 composites. Mat. Res. Bullet. **37**, 1107–1115 (2002)
7. O. Sahin, I. Demirkol, H. Gocmez et al., Mechanical properties of nanocrystalline tetragonal zirconia stabilized with CaO, MgO and Y_2O_3. Acta Phys. Pol. A **123**, 296–298 (2013)
8. Y. Huang, D. Jiang, J. Zhang et al., Fabrication of transparent lanthanum-doped yttria ceramics by combination of two-step sintering and vacuum sintering. J. Am. Ceram. Soc. **92**, 2883–2887 (2009)

9. S. Sankar, K.G.K. Warrier, Aqueous sol-gel synthesis of $LaPO_4$ nano rods starting from lanthanum chloride precursor. J. Sol-Gel Technol. **58**, 195–200 (2011)
10. S. Sankar, A.N. Raj, C.K. Jyothi et al., Room temperature synthesis of high temperature stable lanthanum phosphate–yttria nano composite. Mat. Res. Bull. **47**, 1835–1837 (2012)
11. R. Kumar, S. Chattopadhyaya, A.R. Dixit, B. Bora, M. Zelenak, J. Foldyna, S. Hloch, P. Hlavacek, J. Scucka, J. Klich, L. Sitek, P. Vilaca, Surface integrity analysis of abrasive water jet-cut surfaces of friction stir welded joints. J. Adv. Manuf. Technol. (2016). https://doi.org/10.1007/s00170-016-8776-0
12. J. Wang, Techniques for enhancing the cutting performance of abrasive waterjets. Key Eng. Mat. **257**(258), 521–526 (2004)
13. L.M. Hlavac, I.M. Hlavacova, V. Geryk, S. Plancar, Investigation of the taper of kerfs cut in steels by AWJ. J. Adv. Manuf. Technol. **77**, 1811–1818 (2015)
14. A. Alberdi, A. Rivero, L.N.L. Lacalle, I. Etxeberria, A. Suarez, Effect of process parameter on the kerf geometry in abrasive water jet milling. J. Adv. Manuf. Technol. **51**, 467–480 (2013)
15. A. Alberdi, T. Artaza, A. Suarez, A. Rivero, F. Girot, An experimental study on abrasive waterjet cutting of CFRP/Ti6Al4V stacks for drilling operations. J. Adv. Manuf. Technol. **86**, 691–704 (2015)
16. D. Ghosh, B. Doloi, Parametric analysis and optimization on abrasive water jet cutting of silicon nitride ceramics. J. Prec. Technol. **5**, 294–311 (2015)
17. L. Chen, T.E. Siorest, W.C.K. Wong, Kerf characteristics in abrasive waterjet cutting of ceramic materials. J. Mach. Tools Manuf. **36**, 1201–1206 (1996)
18. A. Hascalik, U. Caydas, H. Gurun, Effect of traverse speed on abrasive waterjet machining of Ti–6Al–4V alloy. Mater. Des. **28**, 1953–1957 (2007)
19. J. Kopac, P. Krajnik, Robust design of flank milling parameters based on grey-taguchi method. J. Mater. Process. Technol. **191**, 400–403 (2007)
20. S. Wang, SuY Zhang, F. Yang, A key parameter to characterize the kerf profile error generated by abrasive water-jet. J. Adv. Manuf. Technol. (2016). https://doi.org/10.1007/s00170-016-9402-x
21. D.S. Srinivasu, D.A. Axinte, Surface integrity analysis of plain waterjet milled advanced engineering composite materials. Proc. CIRP **13**, 371–376 (2014)
22. F. Chen, X. Miao, Y. Tang, S. Yin, A review on recent advances in machining methods based on abrasive jet polishing (AJP). J. Adv. Manuf. Technol. (2016). https://doi.org/10.1007/s00170-016-9405-7
23. H. Onoda, A. Yoshida, The synthesis and properties of bulk lanthanum phosphates obtained by hydrothermal hot pressing. J. Ceram. Process. Res. **13**, 622–626 (2012)
24. Y. Kim, M.H. Hong, S.H. Lee et al., The effect of yttrium oxide on the sintering behavior and hardness of tungsten. Met. Mat. Int. **12**, 245–248 (2006)
25. K. Rajesh, B. Sivakumar, P. Krishna Pillai et al., Synthesis of nanocrystalline lanthanum phosphate for low temperature densification to monazite ceramics. Mat. Lett. **58**, 1687–1691 (2004)
26. P. Mogilevsky, E.E. Boakye, R.S. Hay, Solid solubility and thermal expansion in a $LaPO_4$–YPO_4 system. J. Am. Ceram. Soc. **90**, 1899–1907 (2007)
27. J. Pirso, M. Viljus, K. Juhani, Three-body abrasive wear of TiC–Ni Mo cermets. Tribol. Int. **43**, 340–346 (2010)
28. A.A. Khan, M.M. Haque, Performance of different abrasive materials during abrasive water jet machining of glass. J. Mat. Process. Technol. **191**, 404–407 (2007)
29. G. Fowler, I.R. Pashby, P.H. Shipway, The effect of particle hardness and shape when abrasive water jet milling titanium alloy Ti6Al4V. Wear **266**, 613–620 (2009)
30. H.H.K. Xu, L. Wei, S. Jahanmir, Grinding force and micro crack density in abrasive machining of silicon nitride. J. Mat. Res. **10**, 3204–3209 (1995)

Laser Micromachining of Engineering Materials—A Review

Nadeem Faisal, Divya Zindani, Kaushik Kumar and Sumit Bhowmik

1 Introduction

Conventional machining has formed the backbone of machining processes in industries for ages. This type of machining is achieved by chip removal process generated as a result of direct contact between the tool piece and surface of the workpiece. Nevertheless, recent advancements in the use of high-strength material, environmental aspects, and exponential evolution of micro-level products have given rise to a newer concept in metal machining. Consequently, nontraditional machining processes have emerged to overcome the complications caused by the conventional processes. Laser micromachining is one such technique which produces intricate shapes with the help of lasers.

Laser basically is a coherent, monochromatic light emission radiation that can spread in a straight line with insignificant divergence and happen in an extensive variety of wavelength (ranging from ultraviolet to infrared). Lasers are widely used in manufacturing, communication, measurement, and medical. Energy density of the laser beam could be altered by varying the wavelength. This property has made the lasers proficient for removing extremely small amount of material and has led to the use of lasers to manufacture very small features in workpiece materials. The production of miniature features (dimensions from 1 to 999 μm) in sheet materials using laser machining is termed as laser micromachining. Laser, an abbreviation for light amplification by stimulated emission of radiation is without a doubt one of the

N. Faisal · K. Kumar
Department of Mechanical Engineering, Birla Institute of Technology, Mesra, India
e-mail: ndmfaisal@gmail.com

D. Zindani (✉) · S. Bhowmik
Department of Mechanical Engineering, National Institute of Technology Silchar, Silchar, India
e-mail: divyazindani@gmail.com

K. Kumar et al. (eds.), *Micro and Nano Machining of Engineering Materials*, Materials Forming, Machining and Tribology,
https://doi.org/10.1007/978-3-319-99900-5_6

best and significant developments of the twentieth century. Its continued advancement has been an exciting chapter in the history of science and technology. Development of engineering materials, intricate shape, and irregular size of workpiece confine the utilization of customary machining techniques. Laser beam machining is a standout amongst the most broadly utilized thermal energy based noncontact compose advanced machining process that may be utilized for all scope of materials. The establishment of the laser was laid by Einstein in 1917 when he initially presented the idea of photon discharge, where a photon interacts with an energized particle or atom and causes the outflow of a second photon having a similar frequency, direction, and phase [1].

The laser beam is generated by providing energy to lasing medium from an external source. The electrons at ground state of lasing medium get excited, which causes electrons to move from a lower energy level to a higher energy level and due to very high instability at higher energy level, it comes back to its ground state within a very small time by emitting a photon. The frequency of this emission is then amplified by using two mirrors one with 100% reflective surface and the other with partially reflective one. Laser beam machining is used to perform various operations such as drilling, cutting, turning, milling, etc. [2].

Micromachining with the help of lasers provides a wider wavelength range, extended pulse (from femtosecond to microsecond), and iterative rates (from single to megahertz pulse). This system transmits the pulsed beam with a power of below 1 kW. It is characterized with high precision, minimal collateral damage, applicability to all materials along with fast, and bulk production.

Laser micromachining techniques such as metal micromachining, ultrashort pulses, and femtosecond pulses are used in microfabrication. Metal micromachining includes microdrilling and micromilling. Ultrashort pulses are used in machining glass and in fabrication of waveguides in silicon. Ultrashort pulses provide high intensified beams which produce minimal thermal deprivation [3]. Femtosecond laser micromachining technique is used in the fabrication of optical devices (resonators, splitters, and waveguides), Shape Memory Alloys (SMA), and micromachining of polyuria aerogel [4]. This solution has high structural properties along with its porous structure, which makes its conventional machining complex.

Laser micromachining has also been instigated in the manufacturing of the microsized MESM parts, precision cutting of glass, small-hole drilling in PCB, and laser cutting of various polymides. Many factors contribute toward the effectiveness of laser machining such as pulse duration, intensity of beam, relative velocity, and the pulse number.

2 Literature Review

Researchers have suggested that any arbitrary surface can be generated by using laser ablation. Ablation refers to the process of removing material from the surface of an object by erosive means. Their study showed that fabrication using the beam path approach and profiled path approach techniques of laser micromachining help to create various surfaces of high quality and precision. The beam path approach is characterized by a pulse of shorter wavelength, of the order of 250 nm, depending upon the material. Profiled path of micromachining uses multiple pulsations, generating improper erosion depth without any significant change in laser profile intensity. Klotzbach et al. focused their study toward the effectiveness of laser micromachining. It focuses on the adaptability of short pulse laser systems in the industrial fields. They are highly effective in operations such as drilling and cutting of ceramic and other metals [5, 6].

Knowles et al. termed ablation as a combination of evaporation and melt expulsion, causing irradiating conditions. These conditions can be molded to facilitate high-quality micro-level drilling and cutting [7]. A short pulse laser intensifies the target area of the surface and increases the material properties to critical limits. Brettschneider et al. presented their study on the implementation of laser micromachining as a metallization tool for microfluidic polymer stack [8]. Microfluidic approach piles up multifaceted fluidic procedures, which can be utilized in the modern industry. Laser micromachining builds metal foils into a pile of polymeric layers which sticks the metallic foil together and joins them. Their study depicted the compatibleness of laser machining with various polymers without any chemical additives. Patel and Patel studied the implementation of laser micromachining as a nontraditional cutting process for aluminum based alloys [9]. Laser cutting is basically a thermal process in which a slot is cut by focused traversing of the laser beam. Their studies of different laser cutting processes have been cited as the future scope of machining. Rejab et al. studied the laser micromachining of microelectromechanical components (MEMS) by using FEM technique to build its elemental models. Their work has provided some useful parameters which could help this process replace the conventional machining methods [10].

Parashar et al. discussed the use of laser processing for improved performance and micro-level component developments [11]. Their work related to short pulsated lasers, micro and nanosecond lasers and ultrafast laser showed their widened implementation. Manjaiah et al. have stated the enactment of femtosecond laser in the machining of Titanium based Shape Memory Alloy (SMA) [12]. Femtosecond laser is characterized by high ablation rate, producing high-quality work, and precised dimension components. Agrawal has termed laser micromachining as a miniaturization technology which enables material processing with a high accuracy and flexibility. He studied the direct wiring, mask projection, and interference techniques which would mechanize the material removal through ablation and etching. His precise work throws a shadow on how ultrafast laser pulse can machine material to produce minimal cracks and improve the material properties [3].

Bian et al. explained the application of femtosecond laser in machining of polyuria aerogels [4]. Polyuria aerogel incorporates high thermal and mechanical and thermal properties. Traditionally diamond saw is used to cut this substance which damages the surface and alters its mechanical properties. Femtosecond laser is a noncontact type of material removal process, hence it minimizes the collateral damage and provides an alternative way to cut the polyuria aerogel. Long et al. has made a compound process of collecting a laser beam from the help of an electrolyte process [13]. In this experiment, stainless steel is etched by Laser-Induced Electrochemical Micromachining (LIECM) with a form of ultraviolet laser beam known as excimer laser beam in solution of NaCl. The process has few advantages like little damage characteristics, high etching-rate, and low temperature. Wavelength 248 nm of excimer laser is very short, pulse breadth is narrow, photon energy is quite great with the power of 108 W/cm^2, divergence angle is small and energy density is large. Thermal effects of 248 nm KrF laser are quite small, energy is very great. One of the effects of laser is inducing electrochemical dissolution, another is directly etching materials. He proved through the experimentation that the single-step method of transferring pattern without a mask is practical, but the etching rate is less. Etching rate could be improved by supplying electrolyte and a close distance in between the cathode and anode.

Akhtar et al. recommends that excimer laser, otherwise called the ultraviolet micromachining, is a dry manufacture procedure that speaks of an incredible option for the rapid prototyping and generation of microwave transmission lines, circuit components, and whole miniaturized hybrid microwave integrated circuits with high accuracy [14]. Microwave circuits and transmission lines help in the miniaturization for increasing the performance at high frequencies having lower power, space necessities. Sen et al. explored the fiber laser removed miniaturized scale grooves on Ti-6Al-4V of 0.11 cm thickness in air condition with numerous parametric mixes, and furthermore to establish out the parametric impacts on the groove geometry as far as depth, surface roughness, and width. Ti-6Al-4V is utilized broadly in the hip and dental inserts for its low cytotoxicity and corrosion protection, biocompatibility, fatigue, and wear-resistance [15]. It goes to the understanding that the parametric impact of the miniaturized scale grooves, different process parameters are considered for the test investigation, for example, Number of passes of 1–8; examining rate of 40–1000 mm/s; Pulse frequency of 50–100 kHz; normal power utilization of 2.5–30 W. Groove geometries are estimated utilizing optical microscope and the surface roughness is estimated utilizing AFM (atomic force microscope).

Shalahim et al. mentioned that increasing demands in microelectromechanical systems (MEMS) fabrication are moving to newer demands in manufacturing technology and industry [16]. Laser micromachining accounts for many technological plus points as compared to traditional methods and technologies, which also includes fabrication of complex shape, rapid prototyping being possible and also offers design flexibility. Laser micromachining of acrylic sheet is simulated with finite element models which are developed specifically for them. The temperature plots which are generated through these simulations are equated and debated. The significant aspect of finite element modeling is to reasonably simulate laser micromachining.

Walker et al. have introduced a new laser micromachining technique for manufacturing high-quality, low-cost waveguide structures for frequencies up to 10 THz [17]. He calculated that with this process, the waveguide components of different height and width can be machined to 1 μm accuracy. Through this test beam pattern measurement on a 2 THz corrugated feed-horn can be made. Slatineanu et al. suggested that material removal machining process can be performed in many ways one of them is based on impact phenomena [18]. Mechanical phenomena of microcracking and microcutting lead to the removal of minute quantities of the workpiece material this is how removal is done.

Mishra et al. gave a review on the Laser Beam Micromachining (LBMM) [19]. An overview is given so as to obtain current scenario of LBMM so as to know its capabilities and constraints. Various research activities are performed related with time in Nano, Pico, and femtosecond. The fundamental understanding of the main parameters involved in LBMM process has been discussed. Shalahim et al. carried out a finite element simulation for virtually studying laser micromachining of acrylic material. After performing FEM, they suggested some of the results which are essential to produce defect-free edges in given processing time [16]. They concluded that for proper simulation, proper material model, perfect thermal properties under processing, and mesh design in finite element modeling is required.

Klotzbach et al. refer to the effects of short pulse laser system in various applications. Because of easy installation, moderate cost, high efficiency, it is having a wide scope of applications [5, 6]. Some of the applications include drilling, cutting, material removal, etc. With the help of novel technique, they were able to join two distinct materials polymer and ceramic. It comprises of two steps, one preparation of ceramic surface and another laser-based partial melting of polymer. Malcolm C. Gower et al. discussed aboutthe industrial applications of laser machining [20]. The use of pulsed laser is being carried out in several industries and has various applications in the field of microvia, catheter hole drilling, and nozzle of ink jet printer. Research deals with the accuracy and effectiveness of laser machining in various fields.

Though material rate by pulsed light sources is being examined from the time of the creation of the lasers [21, 22]. During the year 1982, polymers carved by excimer lasers stimulated broad examinations went for utilizing the procedure for micromachining. In the prevailing years, logical and modern inquiries about in this area and field has multiplied to a stunning degree, presumably stimulated by the surprisingly little highlights that can be carved with little harm to encompassing an area of material [23, 24].

Lately, fabricating industry has watched a fast increment demand for small-scale items and miniaturized scale segments in numerous mechanical segments including the hardware, optics, therapeutic, biotechnology, and car parts. These micro-system-based products are an important contributor to a sustainable economy. The laser pulses used in micromachining processes are divided into two groups, one is short (nanosecond) laser pulse and other is ultrashort (picoseconds, femtosecond) laser pulse. Because of the short pulse duration, peak forces of more than 15 GW can be achieved, which offers access to advance removal components, as multiphoton ionization. Because of the short association time, just the electrons inside the material are warmed amid the pulse duration and heat-influenced zones are immaterial [25].

Laser micromachining or ablation phenomenon can happen amid laser—polymer association in two different systems: one of them is photothermal and the other one is photochemical removal. Since polymers display solid absorption in ultraviolet infrared wavelengths, however, powerless absorption at visible and nearly-infrared spectra, in this way, the removal system is a blend of photochemical and photothermal procedures. The compound obligations of the polymer material decay by the photon vitality of the laser light, though in photothermal mechanism, polymer removal happens by quick liquefying and vaporizing. For photochemical removal to happen, the energy of the photons at that wavelength ought to surpass the intermolecular bond energies of the polymer [26]. The photon energy diminishes as the wavelength increments. In this manner, the high vitality of the UV photon breaks the atomic bonds and results in direct photochemical ablation.

The excimer laser has wavelengths accessible at 308 and 248 nm when utilizing gas blends of XeCl and KrF separately. The frequency changed over Nd: YAG lasers with a major wavelength of 1,064 nm have wavelengths of 355 and 266 nm for the third and fourth harmonics, individually. Pulsed Nd: YAG laser beam were utilized by Lau et al. for experimentation to see the impact of HAZ on 2.5 mm thick carbon fiber composite plate with a few parameters [27]. They found that HAZ increments with increment in pulse width, pulse energy, and pulse frequency and diminishing with the feed rate. They likewise watched that heat-influenced zone will be greater when compacted air utilized as assisted gas while argon utilized as assisted gas have smoother cut surface and less HAZ.

As opposed to the UV laser, a CO_2 laser emanates infrared radiation at a wavelength of 10.6 μm which implies that the laser bar dependably removes the basic material photothermally. The region in which focused laser pillar meets the workpiece surface, temperature of the lighted spot rises so quickly that the material first melts and after that decomposes, leaving a void in the workpiece. Laser power, pulse length, cutting rate was utilized as fundamental parameters by [28] for cutting mild steel cutting utilizing beat Nd: YAG and CO_2 laser to perform tests. It was seen by their examination that the CO_2 laser can cut quicker than Nd: YAG, however, Nd: YAG laser serves better surface roughness. It has additionally been watched that the two parameters, i.e., pulse length and cutting rate likewise influences the surface roughness.

3 Types of Lasers Used in LBMM

There is a range of industrial lasers available in a present scenario for micromachining applications. Generally, two types of laser are used for micromachining metals—short pulse laser which emits short pulses of light, of the order of picoseconds to nanoseconds and ultrashort pulse laser which emits ultrashort pulses of light, of the order of femtosecond to ten picoseconds [29]. These lasers are so-called based on the duration of their beam pulses. Consider an example, the pulse produced by a femtosecond laser only lasts femtosecond (a femtosecond is one-millionth of a nanosecond or 10–15 of a second). Similarly, each pulse emitted by a picoseconds laser lasts picoseconds and each pulse released by nanosecond laser lasts nanoseconds. The short pulse means the energy is localized at small depth.

The laser beam can be demonstrated as a radiant energy source with a self-assertive spatial and temporal intensity distribution. The fraction of aggregate beam energy absorbed at the erosion front relies upon the radiation properties of the workpiece and the geometry of erosion front. The power of the incident beam is expressed by Io. The decrement in the laser intensity with the depth is given by Eq. (1)

$$I_x = I_o e^{-\alpha x} \quad (1)$$

where α is optical absorptivity of the material and x is depth into the material. The optical absorptivity (α) of the material records for the decay of laser intensity with depth inside the material. The absorption coefficient relies upon the temperature and wavelength but at constant α, decay of laser intensity with depth is given by Beer–Lambert Law as Eq. (2)

$$I(z) = I_o e^{\alpha z} \quad (2)$$

where I_o is the intensity mainly inside the surface after considering reflection losses. Lambert's law states that absorption of a particular material sample is directly proportional to its thickness (path length). The depth to which the intensity of the laser drops to $1/e$ value of its initial value at the interface is its optical penetration or absorption depth (δ) given by $\delta = 1/\alpha$ [19].

4 Methodology and Mechanism

Laser micromachining takes place with the help of ablation technique. Ablation means the removal of material either thermally or using photochemical erosion [15]. In photochemical or nonthermal ablation, the energy coming out from the incident photon results in the directly breaking of the bond of the molecular chains in the organic materials removal with the help of molecular fragmentation without any thermal damage. So the photon energy during thermal ablation, the excitation energy rapidly converts into heat, causing temperature increase (Fig. 1). Thus, the temperature increase can result in the ablation of material by surface vaporization or spallation which is caused due to thermal stresses [3, 10]. A thermal ablation mechanism is much more useful for the material removal during micromachining of both ceramics and metals. Absorptive coefficients and thermal diffusivity determine the ease of ablation. The large value of absorption coefficient and low value of thermal diffusivity increases ablation efficiency. The ablation of material by internment of laser energy can be aided by using shorter pulses. For longer pulses,

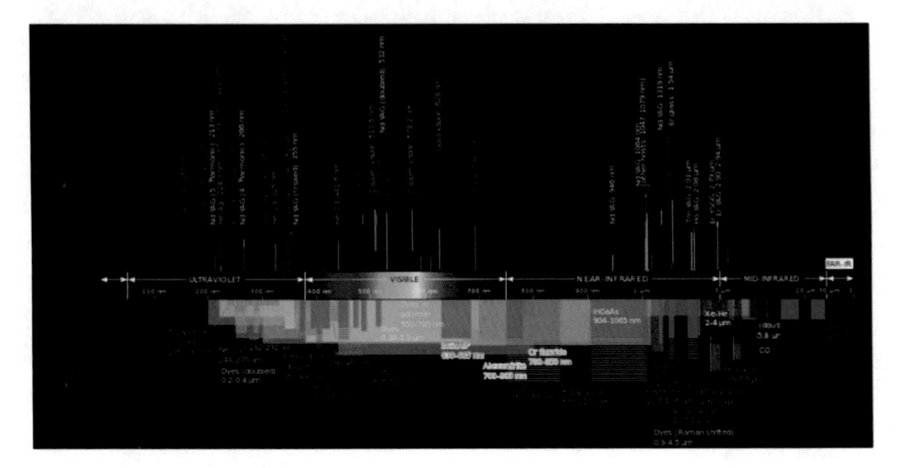

Fig. 1 Schematic diagram of laser micromachining

the absorbed energy will be released to the surrounding by transfer of heat. The ablation process is described by the ablation threshold. Ablation threshold differs from material to material. Laser parameters such as wavelength, eloquence, density of pulse, and material properties govern the ablation process. Excimer lasers and femtosecond lasers feature significantly in power sourcing the machining process. The excimer laser is an ultraviolet laser that is used to micro-machine number of materials without the need of heating them. The photon energy generated by excimer lasers and that of the molecular plastic bonds are comparable. Hence, they are used for plastic and similar metal and not for any metal. When excessive power is applied, the removal phenomenon includes a combination of heating and photon attack. Femtosecond lasers have high power and short pulse duration. These methods generate diminutive heat affected layer which leads to machining of micro-shapes with high precision and lesser defects [3].

Laser ablation is the process of removal of material from a solid surface by irradiating it with a laser beam. At lower energy density, no material removal takes place till a point is reached where material removal starts to happen that is called the ablation threshold [30]. A noteworthy removal of materials happens over a specific threshold power density, and the ejected material structures a luminous ablation plume. This threshold power density required to shape plasma relies upon the retention properties of the given material, the wavelength of laser utilized, and pulse duration [31]. The removed material is directed toward a substrate where it recondense to form a film. The total mass ablated from the target material per laser pulse is called as ablation rate. To improve the reactivity of the background gas with the removed species, either a RF-plasma source or a gas pulse setup are utilized [31].

5 Applications of LBMM

The pulsed lasers are being used for microprocessing materials in several engineering industries. Biomedical catheter hole drilling, microelectromechanical system (MEMS), microvia ink jet printer head, and thin-film scribing are some of the important applications of LASER micromachining.

5.1 Hole Drilling

The capacity of drilling small holes to 1 μm width is a key empowering innovation to fabricate cutting-edge items. Laser micromachining provides answers for key issues in assembling incorporated circuits, interconnects, PC peripherals, hard disks, and telecommunication. The prerequisite for material handling with micron or submicron determination at high speed and low-unit cost can be satisfied by this innovation. The combination for high-determination, precision, speed, and adaptability is permitting laser the growing importance. The combination of high-determination, precision, speed, and adaptability gave laser micromachining to pick up acknowledgment in numerous industries.

5.2 Microvia Hole Drilling in Circuit Interconnection

Microvia drilling in printed circuit boards and adaptable circuits is a gigantic business with several lasers (for the most part in Asia) drilling billions of microvias yearly. A microvia is a little opening bored for electrical conductivity in printed circuit boards. As diameter decreases (below 100 μm), lasers are preferred over mechanical drills. Since copper is very reflective along these lines, Q-switched Nd: YAG lasers are utilized to penetrate the metal, though CO_2 lasers are used to drill the dielectric material. Drilling microvias by removal were initially examined during the mid-1980s utilizing pulsed Nd: YAG and CO_2 lasers [12, 32]. Excimer lasers drove the path in applying it to volume creation when Nixdorf PC plant presented polyimide ablative drilling of 80 μm dia across vias in MCM's—were used to interface silicon chips together in PCs [29]. Other centralized server PC makers, for example, IBM quickly took after suite and introduced their own special generation lines for this application.

Stage 1. Trepanned hole in top copper conductive layer by Nd laser.
Stage 2. Fiber-reinforced composite drilling by CO_2 laser [27, 33].

With less process ventures than different techniques, laser drilling is viewed as the most flexible, solid, and high return innovation for making microvias in thin-film packages.

5.3 Inkjet Printers Nozzle Drilling

It contains a line of little-tapered holes by which ink droplets are squirted into paper. By at the same time diminishing the nozzle dia across and by diminishing the hole pitch and extending the head, increased printer quality can be accomplished. Current printers like HP's Desk Jet 800C and 1600C have nozzle of measurement equivalents to 28 μm giving a resolution of 600 dots for every inch (dpi). At average yields of over 99%, excimer laser mask projection is currently routinely utilized for drilling varieties of nozzles each having indistinguishable size and divider point [34]. The greater part of the ink jet printer heads as of now sold in Asia and US are excimer laser penetrated.

5.4 Biomedical Devices

Lasers make a huge part in the manufacturing of disposable medical devices since the market for these is immense. Microdrilling by excimer lasers is utilized for delicate tests for analyzing Arterial Blood Gases (ABG's) [35]. These comprise of spiral rectangular holes of 50 × 15 μm machined in a 100 μm dia across acrylic (PMMA) optical fiber by a laser. The clean cutting capacity of the laser gives the essential quality that counteract sinking and blockage when embedded into the artery. In intensive care units, choices on patient's ventilator conditions and the organization of various medications are made based on ABG comes about.

5.5 Hole Drilling in Aircraft Engine Components

Jet engine motors which are used nowadays, have up to a huge number of holes bored into different parts, for example, turbine blades, combustion chambers, and nozzle guide vanes. These gaps are under 1 mm in dia, few are through-holes and few are molded holes. Near infrared lasers (for example, high pulse energy Nd: YAG or fiber lasers) are normally utilized as a part of either a trepanning mode or percussion. Such gaps enable a layer of cooling air to cover the parts; which expands life, diminishes maintenance, and accomplishes prevalent performance qualities.

6 Numerical Analysis of Laser Microdrilling

Laser drilling is an extensively used process for fabrication of microvias and micro nozzles as discussed above. After being exposed to short pulses, material is removed by ablation, leaving a crater. By this means, well-shaped micro-hole can be drilled into the material bulk. It is experimentally observed that time delay between pulses has great influence on the shape and machining quality [36]. There is a threshold time at which abrupt shape degradation occurs. To achieve good machining quality and desired shape, longer time delay is preferred. The effect of the time delay can be explained by coupling of residual heat between successive pulses. In this section, our focus will be on quantitative investigation of accumulation of residual heat.

1. The heat dissipation after laser irradiation will be described by a mathematical model.
2. FEM method will be used for the analysis of temperature at the bottom of the crater ablated by a train of pulses.

In FEM, the whole problem domain is subdivided into simpler parts, called finite elements, and it minimizes an associated error function by using vibrational methods from the calculus of variations to solve the problem [37]. Dissimilar to the possibility that joining a few small straight lines can estimate a bigger circle, FEM incorporates techniques for interfacing a few straightforward elements equations over numerous little subdomains, named finite element components to approximate a more complex equation over a bigger area. The threshold-like time delay has been observed for various lasers, including Nd-YAG laser and femtosecond laser. Through this research optical specification for lasers, especially ultrashort laser, used for micromachining can be found.

6.1 *Laser Direct Writing*

The flexibility offered by laser-based direct-write techniques is one of a kind as it gives an opportunity to include, evacuate, and customize distinctive sorts of materials without physical contact between the tool and the material [38]. Laser pulses used to produce patterns can be manipulated to control the creation, structure and properties of 3-dimensional volumes of materials crosswise over length scales traversing six magnitudes of size, from nanometres to millimetres. Laser direct writing is used to produce 2D and 3D structures with the help of two approaches. One of them is direct etching from PMMA sample and other one is replication from metal insert that has been carved by direct laser etching. The laser beam scans in XY plane through a two-axis galvanometer. A high-speed precision translational stage is used to move the sample in Z-axis. The knowledge gained through studies on laser ablation will be used to attain high resolution and excellent quality of machining. The system will be used to manufacture fine parts used in aircrafts and miniaturized satellites.

6.2 *Submicron/Nano Machining Using Ultrashort Laser*

The most recognizable characteristic of ultrashort laser micromachining is that the ablation threshold is clearly defined [39]. Thus, feature size smaller than the laser spot-size can be achieved by controlling the fluency of the laser pulses. However, at the scale of few micron and submicron the quality of machining, throughput and reliability are not acceptable for industrial applications. In this research we aim at developing an ultrashort laser submicron machining, replacing the costly and complicated beam machining in some applications.

7 Recent Advancements in Laser Micromachining

Various micro-electronic industries have shifted their attention towards miniaturization, producing smaller features and holes working with their materials hence implicating higher tolerances (Fig. 2). Laser micromachining has proven to be an optimal tool in delivering high precision, consistent results, faster throughput, higher yields and lower manufacturing cost. Lasers with shorter wavelength such as Ultraviolet rays results in a low peripheral heating, hence have gained popularity.

Fig. 2 Advancement in laser micromachining

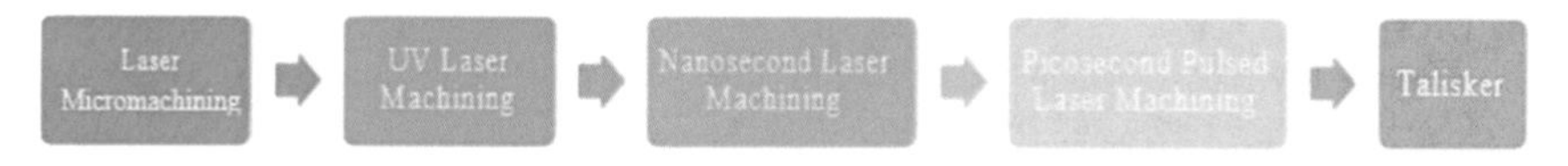

Fig. 3 Nano and pico pulse

Thermal loading of materials can be reduced by using pulsating lasers (Nanosecond laser) (Fig. 3). Nanosecond laser gives high peak power using only modest average power thereby consuming only few watts of overall power. But this method leaves some heart affected zones which may not be good enough for some applications. Picosecond pulsed lasers are being preferred precisely because it does not create any heat zones and produces less heat. Most advanced form of laser machining is the Talisker. It is a two-stage laser which consists of a fiber laser and a free space amplifier. It covers all the drawbacks of previous models and provides fast and accurate machining at an affordable manufacturing cost.

8 Conclusion

Laser micromachining technique has proved to be a sustainable tool in the miniaturization era. Laser ablation is confined to only small areas which absorbs energy. The key for the success of laser ablation is the intensity of pulse, wavelength, and proper shape. Femtosecond and ultrafast lasers produce highly accurate and intricate shaped components. Ultrafast laser pulses produce minimum contamination to the surrounding and fast. Because of its flexible nature laser machining has found wide applications in the fabrication of micro-electronic and mechanical components. They can generally machine any material from metal to glass, ceramics, and many polymers. This study shows a comprehensive study of various laser machining processes and their role in replacing the conventional machining processes. As technology grows, their need will increase. Furthermore, there are certain parameters which can be optimized. The research in the optimization area could further grow the stature of laser machining.

References

1. B.E.A. Saleh, M.C. Teich, *Fundamentals of Photonics*, 2nd edn (Wiley, 2007). https://doi.org/10.1002/9783527635245.ch2
2. S.C. Singh, H. Zeng, C. Guo, W. Cai, Lasers: fundamentals, types, and operations, in *Nanomaterials: Processing and Characterization with Lasers* (2012), pp. 1–34. https://doi.org/10.1002/9783527646821.ch1
3. A.P. Kumar, Laser micromachining: technology and applications. Int. J. Eng. Res. Appl. (IJERA), ISSN: 2248-9622 National Conference on Advances in Engineering and Technology. (AET-29th March 2014)

4. Q. Bian, S. Chen, B.T. Kim, N. Leventis, H. Lu, Z. Chang, S. Lei, Micromachining of polyurea aerogel using femtosecond laser pulses. J. Non-Crystal. Solids **357**(1), 186–193 (2011). https://doi.org/10.1016/j.jnoncrysol.2010.09.037
5. U. Klotzbach, A.F. Lasagni, M. Panzner, V. Franke, Laser Micromachining, in *Fabrication and Characterizationin the Micro-Nano Range, Advanced Structured Materials*, ed. by F.A. Lasagni, A.F. Lasagni, vol. 10 (2011). https://doi.org/10.1007/978-3-642-17782-8_2
6. U. Klotzbach, A.F. Lasagni, M. Panzner, V. Franke, Laser Micromach. Spr. **10**, 29–119 (2011). https://doi.org/10.1007/978-3-642-17782-8
7. H. KnowlesR, Karnakis G. RutterfordM, A.D. Ferguson, Micromachining of metals, ceramics and polymers using nanosecond lasers. Int. J. Adv. Manuf. Technol. **33**, 95–102 (2007). https://doi.org/10.1007/s00170-007-0967-2
8. T. Brettschneider, C. Dorrer, D. Czurratis, R. Zengerle, M. Daub, Laser micromachining as a metallization tool for micro fluidic polymer stacks. J. Micromech. Micro Eng. **23**(3) (2013). https://doi.org/10.1088/0960-1317/23/3/035020
9. R.P. Patel, D.M. Patel, Grey relational analysis based optimization of laser cutting process parameters for aluminum alloy—a review. Int. J. Eng. Res. Technol. (IJERT) **3**(3) (2014). ISSN: 2278-0181
10. M.R.M. Rejab, T.T. Mon, M.F.F. Rashid, N.S.M. Shalahim, M.F. Ismail, Virtual laser-micromachining of MEMS components. Int. J. Recent Trend. Eng. **1**(5) (2009)
11. A. Parashar, J.S. Mann, A. Shah, N.R. Sivakumar, Numerical and experimental study of interference based micromachining of stainless steel. JLMN-J. Laser Micro/ Nano-eng. **4**(2) (2009)
12. M. Manjaiah, S. Narendranath, S. Basavarajappa, Review on non-conventional machining of shape memory alloys. Trans. Nonferrous Metals Soc. China (English Edition) **24**(1), 12–21 (2014). https://doi.org/10.1016/S1003-6326(14)63022-3
13. Y. Long, Q. Liu, Z. Zhong, L. Xiong, T. Shi, Experimental study on the processes of laser-enhanced electrochemical micromachining stainless steel. Optik **126**(19), 1826–1829 (2015). https://doi.org/10.1016/j.ijleo.2015.05.019
14. S.N. Akhtar, S.A. Ramakrishna, J. Ramkumarv, Excimer laser micromachining for miniaturized hybrid microwave integrated circuits. Directions **15**(1) (2015)
15. A. Sen, B. Doloi, B. Bhattacharyya, Experimental studies on fibre laser micro-machining of Ti-6al-4v, in *5th International & 26th All India Manufacturing Technology, Design and Research conference (AIMTDR 2014)*, 14 Dec 2014
16. N.S.M. Shalahim, T.T. Mon, M.F. Ismail, M.F.F. Rashid, M.R.M. Rejab, Finite Element Simulation of Laser-Micromachining". *Proceedings of the International Multi Conference of Engineers and Computer Scientists, IMECS 2010*, vol. 3 (Hong Kong, 17–19 March 2010)
17. C.K. Walker, G. Narayanan, H. Knoepfle, J. Capara, J. Glenn, A. Hungerford, T.M. Bloomstein, S.T. Palmacci, M.B. Stern, J.E. Curtin, Laser micromachining of silicon: a new technique for fabricating high quality terahertz waveguide components, in *Proceedings of 8th international symposium on space terahertz technology* (Harvard University, 1997), p. 358
18. L. Slatineanu, M. Coteaţă, O. Dodun, A. Iosub, L. Apetrei, Impact phenomena in the case of some non-traditional machining processes, in *Project No. ID 625 National Council of Scientific Research in Higher Education (Romania)* (2008)
19. S. Mishra, V. Yadava, Laser beam micromachining (LBMM)—a review. Opt. Lasers Eng. (2015). https://doi.org/10.1016/j.optlaseng.2015.03.017
20. M.C. Gower, Industrial applications of laser micromachining. Opt. Exp. **7**(2), 56–67 (2000). https://doi.org/10.1364/OE.7.000056
21. N. Bloembergen, Laser-material interactions; fundamentals and applications, in *AIP Conference Proceedings*, vol. 288. (1993), pp. 3–10. https://doi.org/10.1063/1.44887
22. J.C. Miller, History, scope and the future of laser ablation, *in Laser Ablation, Principles and Applications*, ed. by J.C. Miller. Springer (1994)
23. Y. Kawamura, K. Toyoda, S. Namba, Effective deep ultraviolet photoetching of polymethyl methacrylate by an excimer laser. Appl. Phys. Lett. **40**(5), 374–375 (1982). https://doi.org/10.1063/1.93108

24. R. Srinivasan, V. Mayne-Banton, Self-developing photoetching of poly (ethylene terephthalate) films by far-ultraviolet excimer laser radiation. Appl. Phys. Lett. **41**(6), 576–578 (1982). https://doi.org/10.1063/1.93601
25. B. Wilhelmi, J. Herrmann, *Lasers for Ultrashort Light Pulses,* OSTI-Identifier:5733239 (United States, 1987)
26. M.S. Brown, C.B. Arnold, Fundamentals of laser-material interaction and application to multiscale surface modificatication. Springer Ser. Mat. Sci. **135**(0933–033X), 91–120 (2010). https://doi.org/10.1007/978-3-642-10523-4.
27. J.R. Lankard, G.E. Wolbold, Laser ablation of polyimide in a manufacturing facility. Appl. Phys. A54, 355 (1992)
28. F.O. Olsen, L. Alting, Pulsed laser materials processing, ND-YAG versus CO_2 Lasers. CIRP Ann. Manuf. Technol. **44**(1), 141–145 (1995). https://doi.org/10.1016/S0007-8506(07)62293-8
29. J. Meijer, Laser beam machining (LBM), state of the art and new opportunities. J. Mat. Process. Technol. **149**, 2–17 (2004). https://doi.org/10.1016/j.jmatprotec.2004.02.003
30. X. Liu, D. Du, G. Mourou, Laser ablation and micromachining with ultrashort laser pulses. IEEE J. Quant. Electronics, 33(10), 1706–1716 (1997). https://doi.org/10.1109/3.631270
31. E. Ohmura, I. Fukumoto, Study on fusing and evaporating process of fcc metal due to laser irradiation using molecular dynamics. Int. J. Jpn. Soc. Precis. Eng. **30**, 47–48 (1996)
32. F.J. McClung, R.W. Hellwarth, Giant optical pulsations from ruby. J. Appl. Phys. **33**(3), 828–829 (1962). https://doi.org/10.1063/1.1777174
33. R.S. Patel, T.F. Redmond, C. Tessler, D. Tudryn, D. Pulaski, Production benefits from excimer laser tools, in *Laser Focus World* (1996)
34. C. Rowan, Excimer lasers drill precise holes with higher yields, in *Laser Focus World* (1995)
35. M.C. Gower, Excimer lasers for surgery and biomedical fabrication, in *Nanotechnology in Medicine and the Biosciences*, ed. by R.R.H. Coombs, D.W. Robinson (Gordon & Breach, New York, 1996)
36. B. Tan, K. Venkatakrishnan, Thermal coupling in multishot laser microvia drilling for interconnection application. J. Vac. Sci. Technol. B Microelectron. Nanometer Struct. **24**, 211–215 (2006). https://doi.org/10.1116/1.2162573
37. J.N. Reddy, *An Introduction to the Finite Element Method* (McGraw-Hill, New York, USA, 2006). ISBN 9780072466850
38. C.B. Arnold, A. Piqué, Laser Direct-Write Processing. MRS Bulletin **32**, 15 (2007)
39. K. Venkatakrishnan, B. Tan, N.R. Sivakumar, Submicron machining of metallic film by low influence ultrashort". Opt. Laser Technol. **34**, 575–578 (2002)
40. S.D. Allen, M. Bass, M.L. Teisniger, Comparison of pulsed Nd:YAG and pulsed CO_2 lasers for hole drilling in printed circuit board materials, in *CLEO Conference Summary* (1982)
41. F. Bachman, Excimer lasers in a fabrication line for a highly integrated printed circuit board. Chemtronics **4**, 149 (1989)
42. P. Crosby, Get to know types of lasers, in *Materials Processing Units from Coherent Inc.* June 2002.
43. J.P. Desbiens, P. Masson, ArF excimer laser micromachining of Pyrex, SiC and PZT for rapid prototyping of MEMS components. Sens. Actuators A Phys. **136**(2), 554–563 (2007). https://doi.org/10.1016/j.sna.2007.01.002
44. W.S. Lau, W.B. Lee, S.Q. Pang, Pulsed Nd: YAG laser cutting of carbon fibre composite materials. CIRP Ann. Manuf. Technol. **39**(1), 179–182 (1990). https://doi.org/10.1016/S0007-8506(07)61030-0
45. J. Meijer, K. Du, A. Gillner, D. Hoffmann, V.S. Kovalenko, T. Masuzawa, W. Schulz, Laser machining by short and ultrashort pulses, state of the art and new opportunities in the age of the photons. CIRP Ann. Manuf. Technol. **51**(2), 531–550 (2002). https://doi.org/10.1016/S0007-8506(07)61699-0
46. H.W. Mocker, R.J. Collins, Mode competition and self-locking effects in a q-switched ruby laser. Appl. Phys. Lett. **7**(10), 270–273 (1965). https://doi.org/10.1063/1.1754253

47. E. Ohmura, I. Fukumoto, I. Miyamoto, Molecular dynamics simulationon laser ablation and thermal shock phenomena, in *Proceedings of the ICALEO* (1998), pp. A45–A54
48. R. Paschotta, R.P. Photonics, C. Gmbh, Solid state lasers for ultrashort pulses—a diverse family, in *Photonick International* (2006), pp. 1–4
49. S. Ronald, *Fundamentals of Laser Micromachining* (CRC Press, A Taylor & Francis Group, USA, 2012). ISBN 9781439860557
50. M.N. Watson, Laser drilling of printed circuit boards, in *Circuit World,* (1984), pp. 11, 13

Experimental Analysis of Wire EDM Process Parameters for Micromachining of High Carbon High Chromium Steel by Using MOORA Technique

Sarat Kumar Sahoo, Sunita Singh Naik and Jaydev Rana

1 Introduction

Due to modernization in mechanical industry, various nontraditional methods have been developed. Wire electric discharge machining process (WEDM) is one type of EDM process under unconventional machining processes for preparation of intricate contours and profiles in any conducting material with very high accuracy. WEDM process is broadly used in electronics, medical, aerospace, nuclear, automobile, mold and die making industries, etc. This is a thermoelectric material removal method, where workpiece is eroded by a sequence of discrete electric sparks stuck between the wire electrode and workpiece submerged in a dielectric fluid medium. The electrical discharge generates high amount of heat, which is sufficient to melt and vaporize tiny amount of material from the surface workpiece. These evaporated materials are cooled by dielectric and form tiny particles called debris, which is removed by the flushing of dielectric fluid. Since no cutting forces are applied, WEDM process is ideal for delicate parts without any mechanical stresses. It is possible to make tapered parts with different profiles at the top and bottom as the movement of wire can be inclined to any direction by wire guide. Working principle of Wire EDM method is shown in Fig. 1.

Scott et al. [1] studied and developed model to predict output parameters in WEDM process. Miller et al. [2] applied regression analysis to find the effect of i/p parameters on o/p by using four different types of w/p material. Kuriakose and Shunmugam [3] applied genetic algorithm for optimization of WEDM o/p

S. K. Sahoo (✉)
Department of Mechanical Engineering, CVR College of Engineering, Hyderabad 501510, Telangana, India
e-mail: saratkumar222@gmail.com

S. S. Naik · J. Rana
Department of Mechanical Engineering, Veer Surendra Sai University of Technology (VSSUT), Burla, Sambalpur 768018, Odisha, India

K. Kumar et al. (eds.), *Micro and Nano Machining of Engineering Materials*, Materials Forming, Machining and Tribology,
https://doi.org/10.1007/978-3-319-99900-5_7

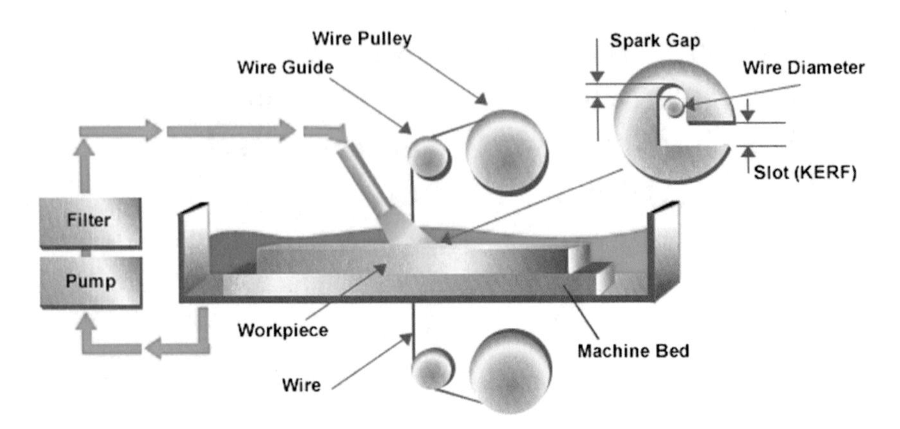

Fig. 1 Working principle of WEDM process

parameters. Ramakrishnan and Karunamoorthy [4] optimized parameters by the application of Taguchi methodology. Kumar and Agarwal [5] optimized MRR and SR by the help of genetic algorithm technique. Fard et al. [6] investigated on dry WEDM of MMC using ANFIS technique. Pragadish and Kumar [7] adopted GRA and ANOVA for optimization of dry EDM parameters on D2 steel material. Nain et al. [8] optimized wire-cut EDM of super alloy metal by using L_{27}OA and GRA methodology.

Multi-objective optimization by ratio analysis (MOORA) technique was first introduced in 2004 by Brauers [9]. This method has been successfully applied in various engineering/management fields, for example, selection of best intelligent manufacturing system [10], parametric optimization of milling process [11], system for selection of untraditional machine [12], etc. Gadakh et al. [13] used MOORA technique for welding optimization problem. Chaturvedi and Sharma [14] used MOORA optimization technique to optimize the Wire EDM material removal rate for OHNS steel.

From the above literatures, it is found that the Wire EDM is a very important process in various manufacturing activities. A number of researchers have tried to increase the working characteristics, viz., surface finish, material removal rate, cutting speed, dimensional accuracy, kerf width, white layer, etc., in Wire EDM method. But the complete application of WEDM process is not entirely resolved due to its complex nature and involvement number of parameters in this process.

Very few researchers have applied MOORA technique for optimization of WEDM process, which motivated the author to use this optimization technique in this current work. And also, few researches are carried out to optimize wire feed rate and calculation of kerf width. This is taken into consideration in this experiment.

2 Methods and Material

2.1 MOORA Methodology

The multi-response optimization is a technique for optimization of two or more contradictory attributes simultaneously. Among the criteria, some may be maximized and some may be minimized. MOOR technique considers both maximization and minimization principles for the selection of desired alternatives from all the alternatives.

In MOORA technique, first step is to develop a decision matrix showing the pertinent evaluation attributes for different alternatives.

$$X = \begin{bmatrix} x_{11} & x_{12} & \cdots & x_{1b} \\ x_{21} & x_{22} & \cdots & x_{2b} \\ \cdots & \cdots & \cdots & \cdots \\ \cdots & \cdots & \cdots & \cdots \\ x_{a1} & x_{a2} & \cdots & x_{ab} \end{bmatrix} \quad (1)$$

Here, x_{ij} is the performance measure of ith alternative in jth criterion, wherein "a" is number of alternatives and "b" is number of criteria.

The second step is the normalization of decision matrix. Brauers et al. proposed a ratio system to calculate the normalized performance as shown in Eq. 2.

$$x_{ij}^{*} = x_{ij} / \sqrt{\sum\nolimits_{i=1}^{a} x_{ij}^{2}} \quad (2)$$

Here, x_{ij}^{*} lies in [0, 1] and represents the normalized performance.

The third step is the calculation of the overall performance as per Eq. 3.

$$y_A = \sum_{j=1}^{k} x_{ij}^{*} - \sum_{j=k+1}^{b} x_{ij}^{*} \quad (3)$$

Here, "k" is the number of parameters that needs to be maximized and "b − k" is the number of parameters that needs to be minimized. "y_A" is the assessment value of ith alternative with respect to all the parameters.

In many situations, it is witnessed that certain o/p parameters are more essential than other o/p parameters. To provide more importance to certain parameters, individual weightage value needs to be multiplied. By considering the weightage value, Eq. 3 can be modified as per Eq. 4.

$$y_A = \sum_{j=1}^{k} w_j \times x_{ij}^{*} - \sum_{j=k+1}^{b} w_j \times x_{ij}^{*} \quad (4)$$

Subjected to

$$\sum_{j=1}^{n} W_j = 1 \quad (5)$$

Here, W_j is the weightage value of o/p parameters. The value of "y_A" can be positive or negative. Ranking of y_A value shows the sequence of preference. The highest y_A value shows best sequence of operation and the lowest y_A value shows the worst sequence.

2.2 Analysis of Variance

ANOVA is a general arithmetical technique to find out the percentage of contribution of each factor/input parameter on the overall result of the experiment.

These are the steps followed for the determination of ANOVA table:

Step 1: Determination of total sum of square (SS_t)

$$SS_t = \sum_{i=1}^{m} (\eta_j - \bar{\eta})^2 \quad (6)$$

Step 2: Determination of sum of square due to factor (SS_f)

$$SS_f = q \times \sum_{q=1}^{q} (\eta_j - \bar{\eta})^2 \quad (7)$$

Step 3: Determination of sum of square due to error (SS_e)

$$SS_e = SS_t - \sum_{q=1}^{q} SS_f \quad (8)$$

Step 4: Determination of degrees of freedom (DOF) for any factor

$$\text{DOF} = \text{No. of level} - 1 \quad (9)$$

Step 5: Determination of total degrees of freedom (DOF)

$$\text{Total DOF} = \text{Total no. of expt.} - 1 \quad (10)$$

Step 6: Determination of mean square for each factor (MS_f)

$$MS_f = \frac{SS_f}{DOF} \quad (11)$$

Step 7: Determination of percentage of contribution of each factor

$$\%Contribution = \frac{SS_f}{SS_t} \times 100 \quad (12)$$

3 Material

High-carbon and high-chromium (HCHCr) steel possesses high wear out resistance, high dimensional stability, high compressive strength, and deep hardening properties. It has the properties of low specific heat and tendency to get strain hardened. The hardness value of this material is very high. So it cannot be easily machined because of large cutting force and large power consumption required during machining that may also result in rapid tool wear and tool failure. HCHCr is generally used in thread rolling dies, punches, draw plates, hobs, cold extrusion tools and dies, piercing, cutters, gauging apparatuses, pressure casting molds, reamer, etc. Keeping in application point of view, HCHCr of grade D3 type has been selected for this experiment. The chemical composition of the workpiece material is shown in Table 1.

4 Experimentation

Experiments have been performed on Electronica ecocut (Elpuls 15 pulse generator) CNC wire electrical discharge machine to study the rate of material removal, kerf width, and surface roughness by changing the pulse duration time, pulse off time, and wire electrode feed rate. HCHCr steel of length 100 mm, width 20 mm, thickness 10 mm, and weight 287 gm has been taken as workpiece material. Brass wire electrode of 0.25 mm diameter has been taken for cutting. Demineralized water is taken as dielectric fluid. Figure 2 shows the clear view of WEDM machine and Fig. 3 shows the experimental setup in WEDM machine.

Table 1 Chemical composition of HCHCr steel (wt%)

Carbon (C)	Chromium (Cr)	Silicon (Si)	Manganese (Mn)	Iron (Fe)
2.01	12.21	0.28	0.25	Remainder

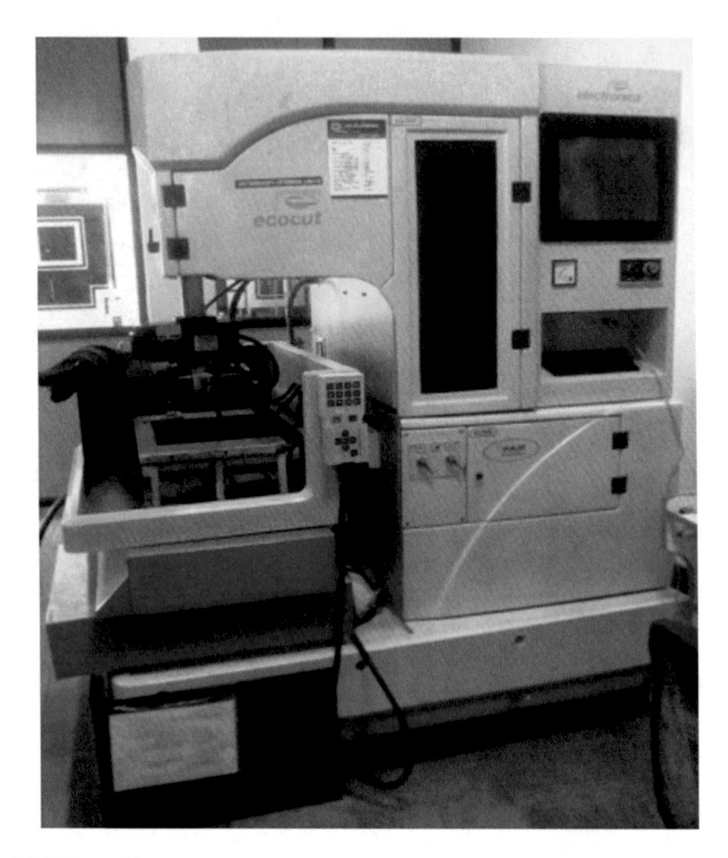

Fig. 2 WEDM machine

Taguchi's L_9 orthogonal array of three factors and three levels is used for the experiment. Input parameters are pulse width time (15, 20, and 25 μs), pulse off time (25, 30, and 35 μs), and wire feed rate (8, 10, and 12 m/min). Servo voltage (15 V), flushing pressure, current, etc., are taken as fixed parameters. Table 2 shows the experimental value of response variables.

During the experiment, MRR, KW, and average SR have been calculated by measuring the initial weight, initial length, final weight, final length, and time required for cutting. After cutting all specimens, surface roughness is measured for each cutting surface. For identification of each cut, the workpiece has been numbered for each cutting sample.

Material removal rate is measured in g/min. For this present experiment, the weight has been measured by digital weighing machine (readability 0.1 gm) that can measure 0.1 g. MRR has been determined by dividing machining time with difference of initial and final weights of material in each cut.

Fig. 3 Experimental setup

Table 2 Experimental value of response variables

Expt. no.	T_{ON} (μs)	T_{OFF} (μs)	WF (m/min)	MRR (g/min)	KW (mm)	Avg. SR (μm)
1	15	25	8	0.047	0.35	1.421
2	15	30	10	0.049	0.34	1.402
3	15	35	12	0.050	0.34	1.397
4	20	25	10	0.058	0.39	1.844
5	20	30	12	0.061	0.38	1.836
6	20	35	8	0.055	0.36	1.749
7	25	25	12	0.072	0.43	2.321
8	25	30	8	0.067	0.40	2.105
9	25	35	10	0.071	0.39	1.972

$$\text{MRR} = (\text{Initial weight of W/P} - \text{Final weight of W/P})/\text{machining time} \tag{13}$$

Kerf width can be calculated by measuring the length of w/p before and after each cutting operation. In this present experiment, the length has been measured by using digital Vernier scale that can measure 0.01 mm.

In this work, the surface roughness of each specimen has been measured by Surtronic 25 roughness tester.

5 Result and Discussion

The effects of input factors on MRR can be found out by using Minitab software as shown in Fig. 4. From the graph, it is found that due to increase in pulse duration time, MRR increases in a rapid rate because of longer spark. Increase in pulse off time MRR decreases in a very slower rate. Increase in wire feed rate increases MRR but in a slower rate than the effect of pulse on time.

The effects of input parameters on KW can be found out by using Minitab software as shown in Fig. 5. From the graph, it is found that due to increase in the pulse duration time, the kerf width increases in a very rapid rate. Increase in pulse off time decreases the kerf width in a slower rate. Increase in rate of wire feed in turn increases the kerf width in a very slower rate.

The effects of input parameters on SR can also be found out by using Minitab software as shown in Fig. 6. It is clear from the graph that, increase in pulse duration time increases SR in a rapid rate. Increase in pulse off time decreases SR in a slower rate. Due to an increase in wire feed rate, SR initially does not vary more but after a certain limit increases because of higher cutting.

Optimization is used to maximize the desired benefits/outputs by minimizing the efforts/inputs required. In this study, MOORA multi-response optimization technique has been used for optimization of all the conflicting process parameters

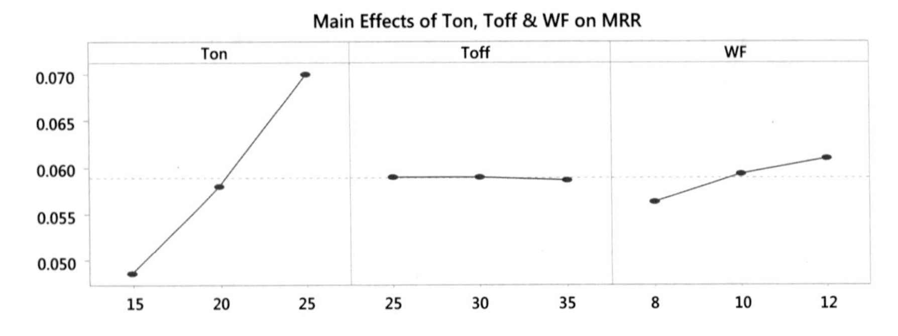

Fig. 4 Main effects of input factors on MRR

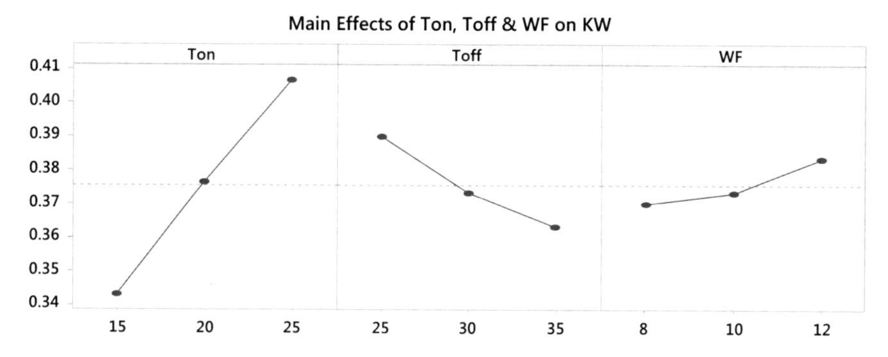

Fig. 5 Main effects of input parameters on KW

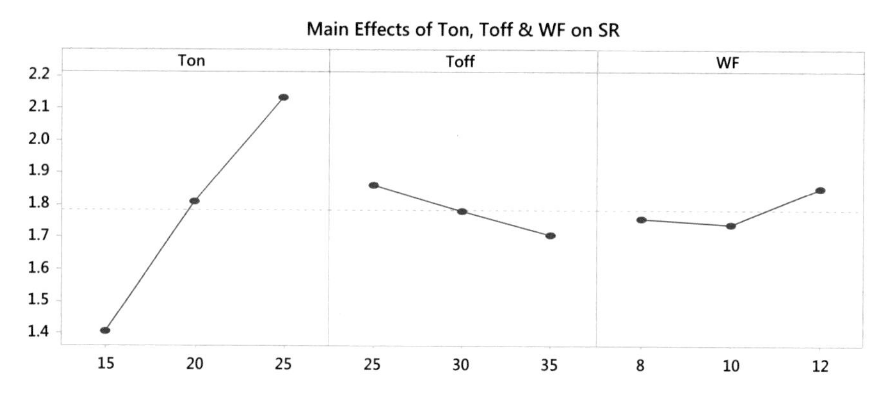

Fig. 6 Main effects of input parameters on SR

simultaneously. Weightage factor can be provided to give more importance to certain parameters (Outputs). Weightage factors are generally chosen according to the importance of different output parameters. In this experiment, weightage factor of individual parameters is not taken into consideration. But according to the requirement of industry, it can be provided for desired benefits. To show the application and calculation of weightage factor, here it is taken the same weightage factor as 0.33 for each parameter. Weightage factor can be changed according to the requirement of the industries. Optimized data values and final ranking are shown in Table 3.

To find the optimum sequence of machining performance, level average value has been calculated as shown in response Table 4. Response table also provides rank as per the most effective parameter.

The highest value of each factor is taken as optimum level for overall performance characteristics. So the optimum level is $(T_{ON})_1$, $(T_{OFF})_3$, and$(WF)_2$. The sequence of highest to lowest order of influence is T_{ON}, T_{OFF}, and WF as shown in

Table 3 Optimized data by MOORA method

Expt. no.	MRR (x_{ij}^*)	KW (x_{ij}^*)	SR (x_{ij}^*)	MRR (V_{ij}^*)	KW (V_{ij}^*)	SR (V_{ij}^*)	Y_A	Rank
1	0.2630	0.3097	0.2617	0.0877	0.1032	0.0872	−0.1028	3
2	0.2742	0.3009	0.2582	0.0914	0.1003	0.0861	−0.0950	2
3	0.2798	0.3009	0.2573	0.0933	0.1003	0.0858	−0.0928	1
4	0.3246	0.3451	0.3397	0.1082	0.1150	0.1132	−0.1201	6
5	0.3414	0.3363	0.3382	0.1138	0.1121	0.1127	−0.1110	5
6	0.3078	0.3186	0.3222	0.1026	0.1062	0.1074	−0.1110	5
7	0.4029	0.3805	0.4275	0.1343	0.1268	0.1425	−0.1351	8
8	0.3749	0.3540	0.3877	0.1250	0.1180	0.1292	−0.1223	7
9	0.3973	0.3451	0.3632	0.1324	0.1150	0.1211	−0.1037	4

Table 4 Response table of MOORA method

Factors	Level-1	Level-2	Level-3	Max–Min	Rank
T_{ON}	−0.09687	−0.11404	−0.12034	0.02347	1
T_{OFF}	−0.11932	−0.10943	−0.10250	0.01682	2
WF	−0.11203	−0.10625	−0.11297	0.00672	3

Table 4. We can also find out the optimum level by graphical method using Minitab software. Figure 7 shows the optimum level machining performance from the assessment value.

Percentage contribution of each input parameter affecting the multiple performance characteristics can be found out by using ANOVA technique shown in Table 5.

ANOVA table shows that the machining performance is very highly influenced by pulse on time (about 60%) and very less influenced by wire feed (about 6%). Machining performance is second largely influenced by pulse off time (about 28%).

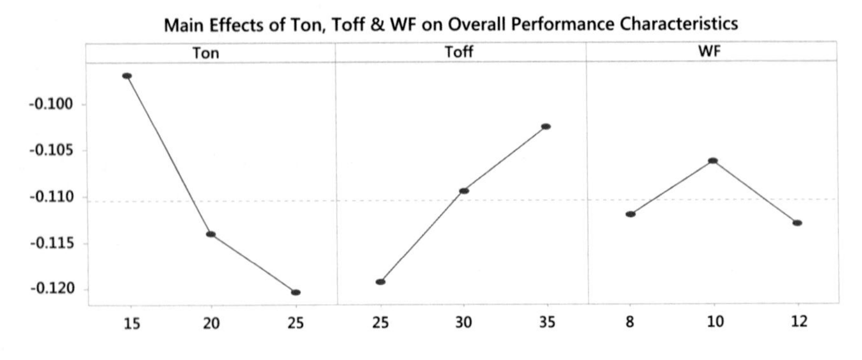

Fig. 7 Optimum level machining performance from Y_A value

Table 5 ANOVA result for overall characteristics

Factors	SS	DOF	MS	% Contribution
T_{ON}	0.000885	2	0.000443	59.27
T_{OFF}	0.000429	2	0.000214	28.73
WF	0.000099	2	0.000040	6.63
Error	0.000079	2	0.000050	5.29
Total	0.001493	8	0.000186	100

So, during machining, we have to give more focus on "pulse length time" than "pulse off time", and "wire feed rate".

This calculated experimental data show best optimum sequence for cutting of high-carbon, high-chromium material by wire-cut EDM process. All the factors investigated in the study have been found very significant on output response.

6 Conclusions

In this study, experiments have been conducted in order to inspect the effects of cutting factors, for example, pulse length time, pulse off time, and rate of wire feed on rate of material removal, kerf width, and surface roughness during cutting by Wire EDM process of high-carbon, high-chromium steel. This research applied MOORA technique effectively for multi-objective optimization with taking importance of different response variables. This paper shows the idea to utilize MOORA technique for various complex processes. We conclude that the high-carbon, high-chromium steel can be machined easily with less time and high accuracy by Wire EDM process. It is found that MRR, KW, and SR increase rapidly with an increase in pulse on time, whereas MRR, KW, and SR decrease in a slower rate with an increase in pulse off time. And, with an increase in servo voltage; the material removal rate, kerf width, and surface roughness increase.

From the overall effect, it is found that the optimum level is $T_{on1} = 15$ μs, $T_{off3} = 35$ μs, and $WF_2 = 10$ m/min in order to maximize MRR and minimize KW and SR. This result also indicates the best combination of the input parameter in order to get high productivity with high accuracy, which can be used effectively in industries. The percent contribution of pulse length time (60%) and pulse off time (28%) shows their significance on overall performance characteristics. Wire feed rate has less influence on machining. So, during machining on WEDM, we have to give more emphasis on pulse on time.

In future work, servo voltage, peak current, wire material type, flushing pressure, type of dielectric, wire tension, etc., can be taken for investigation. The correlation of cutting parameters and surface texture, like whiter layer, microcrack, and surface topography, can be studied to get higher surface finish and to decrease white layer, which is the major demerit of Wire EDM process.

The main challenge for Wire EDM process is to machine the nonconducting material. In future combination, Wire EDM with other process (hybrid process) can be used for effective machining of semiconducting and nonconducting materials with high accuracy and surface finish.

References

1. D. Scott, S. Boyina, K.P. Rajurkar, Analysis and optimization of parameter combination in wire electrical discharge machining. Int. J. Prod. Res. **29**(11), 2189–2207 (1991)
2. S.F. Miller, A.J. Shih, J. Qu, Investigation of the spark cycle on material removal rate in wire electrical discharge machining of advanced material. Int. J. Mach. Tools Manuf. **44**(4), 391–400 (2004)
3. S. Kuriakose, M.S. Shunmugam, Multi-objective optimization of wire electro discharge machining process by non-dominated sorting genetic algorithm. J. Mater. Process. Technol. **170**(1), 133–141 (2005)
4. R. Ramakrishnan, L. Karunamoorthy, Multi response optimization of wire EDM operations using robust design of experiments. Int. J. Adv. Manuf. Technol. **29**(1), 105–112 (2006)
5. K. Kumar, S. Agarwal, Multi-objective parametric optimization on machining with wire electric discharge machining. Int. J. Adv. Manuf. Technol. **62**(5), 617–633 (2012)
6. R.K. Fard, R.A. Afza, R. Teimouri, Experimental investigation, intelligent modeling and multi-characteristics optimization of dry WEDM process of Al–SiC metal matrix composite. J. Manuf. Process. **15**, 483–494 (2013)
7. N. Pragadish, M. Pradeep Kumar, Optimization of dry EDM process parameters using grey relational analysis. Arab. J. Sci. Eng. **41**, 4383–4390 (2016)
8. S.S. Nain, D. Garg, S. Kumar, Modeling and optimization of process variables of wire-cut electric discharge machining of super alloy Udimet-L605. Eng. Sci. Technol. Int. J. **20**, 247–264 (2017)
9. W.K. Brauers, *Optimization Methods for a Stakeholder Society: A Revolution in Economic Thinking by Multi-Objective Optimization* (Kluwer Academic Publishers, USA, 2004)
10. U.K. Mandal, B. Sarkar, Selection of best intelligent manufacturing system (IMS) under fuzzy MOORA conflicting MCDM environment. Int. J. Emer. Technol. Adv. Eng. **2**(9), 301–310 (2012)
11. V.S. Gadakh, Application of MOORA method for parametric optimization of milling process. Int. J. Appl. Eng. Res. **1**(4), 743–758 (2011)
12. A. Sarkar, S.C. Panja, D. Das, B. Sarkar, Developing an efficient decision support system for non-traditional machine selection: an application of MOORA and MOOSRA. Prod. Manuf. Res. **3**(1), 324–342 (2015)
13. V.S. Gadakh, V.B. Shinde, N.S. Khemnar, Optimization of welding process parameters using MOORA method. Int. J. Adv. Manuf. Technol. **69**(9), 2031–2039 (2013)
14. V. Chaturvedi, A.K. Sharma, Parametric optimization of cnc wire cut edm for ohns steel using MOORA methodology. Int. J. Mech. Prod. Eng. **2**(12), 55–60 (2014)

Index

K. Kumar et al. (eds.), *Micro and Nano Machining of Engineering Materials*, Materials Forming, Machining and Tribology,
https://doi.org/10.1007/978-3-319-99900-5

Printed in the United States
By Bookmasters